NAVY SEAL SNIPER

NAVY SEAL SNIPER

AN INTIMATE LOOK AT THE SNIPER OF THE 21st CENTURY

NAVY SEALS GLEN DOHERTY and
BRANDON WEBB

Foreword by NAVY SEAL CHRIS KYLE
Foreword by DON MANN

Skyhorse Publishing

CONTENTS

FOREWORD

Next to BUD/S (Basic Underwater Demolition/SEAL Training), SEAL sniper training was the hardest school I ever went through. The stress and pressure the instructors put on us during our training was intense—but necessary. There are some great sniper programs in the U.S. military, like the USMC's Scout Sniper Course and the army's Special Operations Target Interdiction Course (SOTIC), but the SEAL Sniper Course is a very different animal. It is one of the world's best programs.

Never has the sniper been more relevant to our military than it is today in the twenty-first century. As I describe at some length in my own book, the threats, environments, and missions modern snipers face in the course of the Global War on Terror are very challenging. And that's putting it mildly.

As Brandon and Glen mention, few books on sniping are written by actual snipers, and it's about damn time that changed. These guys have been through the training and done the work in combat. They do a great job of giving you a glimpse of the sniper's world: history, weapons, gear, great pictures, and a personal perspective. Even better, they tell it in plain and simple English.

In *American Sniper* I shed some light on the sniper community and tell the greater story of sacrifice and teamwork involving *all* warfighters on the battlefield. After all, it's about teamwork and working toward a common objective. The lone sniper can deal a massive blow to the enemy's psyche, but a sniper cannot win the war alone. It takes a team, and these guys understand that.

Enjoy a peek behind the curtain.

—Chris Kyle
Navy SEAL sniper and bestselling
author of *American Sniper*

FOREWORD
TO THE 2016 EDITION

In today's Global War on Terror, snipers are more relevant than in any other time in our nation's history. Glen Doherty, one of the four Americans tragically killed in the Benghazi Consulate attack on September 11, 2012, and Brandon Webb provide a unique and very exciting glimpse of the brutal physical and mental training and missions that todays Navy SEAL Snipers endure.

From the moment I checked into my first SEAL Team, (ST-1), I quickly gained an appreciation of just how hard our SEAL snipers worked, how meticulous they were not only in shooting but in everything we did as SEALs. They were quiet, intense, and always so very professional. Our snipers were as skilled as every other SEAL in all other aspects of the duties of a SEAL but, in addition, they were the snipers for the team, the over-watch and the "eyes from above."

Snipers are routinely tasked with making life or death decisions for unsuspecting men, woman and children. These snipers have to process a great deal of information before making the decision of pressing that trigger back and launching a round that would take the life from another human being.

All SEALs and other warriors are trained for combat, trained to kill, et cetera, but it is somewhat a different story for a sniper when he can clearly see the expressions on his victim's face as he fires his lethal round.

The selection process of becoming a SEAL sniper is incredibly challenging. Basic Underwater Demolition/SEAL (BUD/s) is known as the most difficult military training in the world and produces the finest warriors on the planet. It has always impressed me just how frequently so many highly trained SEALs do not pass the arduous three-month SEAL sniper course.

Selection as a SEAL sniper is based on a multitude of factors including: marksmanship and stalking abilities; proficiency in the arts of stealth and concealment; an exceptionally high degree of mental focus and concentration, intelligence, and decision making abilities; and the mental and physical capacity to crawl through and set up on target for hours or sometimes days through jungle, desert, mountain, arctic, and urban war-zone terrain.

But, to put it all in perspective, the sniper selection and training process does not compare to the realities that a SEAL sniper faces when operating in a war zone. Training is just the beginning. The test is combat. No matter how proficient our SEAL sniper instructors are in making training realistic, there is no way to train a person for actually making the shoot—no-shoot decisions and dealing with the consequences of those decisions.

It is very difficult for most people to comprehend the scope of dealing with the life and death decisions these snipers routinely make. How do they cope with taking the shot without having 100 percent assurance the person was a threat? What happens to the psyche of a sniper after he has killed a person, ten people, or one hundred or more people?

Snipers generally have a similar mindset regarding the taking of lives. Chris Kyle, the "Devil of Ramadi," who served four tours of Iraq and has 160 confirmed kills and 255 probable kills, is the most accomplished sniper in US history. Chris stated that "every person I killed, I strongly believe that they were bad." Chris claimed to have no regrets and referred to the people he killed as "savages." He stated that, "When I do go face God there is going to be lots of things I will have to account for but killing any of those people is not one of them."

In 2003, during Chris's first tour as a sniper, he was tasked with being the over-watch for Marine battalion in a terrorist-rich Iraqi town. As locals gathered around the Marines during their movement, Chris observed a woman with a child approach the Marines. She had a grenade in her hand. This was the first of many life and death decisions Chris faced as a sniper and was his first kill of many more to follow. He was forced to make a quick decision, but in Chris's view, "She made the decision for me, it was either my fellow Americans die or I take her out."

There have been countless studies that show that snipers tend to empathize with their victims as real people and are less likely than other soldiers to dehumanize their enemy. Most soldiers in today's wars refer to the enemy as "terrorists," but snipers generally referred to them as human beings or even as legitimate warriors.

This might be because snipers often see their victims with great visual clarity and often observe them for hours and sometimes days. Although a sniper shot is often made from a distance it can become more personal than killing in typical combat.

"Here is someone whose friends love him and I am sure he is a good person because he does this out of ideology," said a sniper who watched through his scope as a family screamed and cried over the man he had just killed. "But we from our side have prevented the killing of innocents, so we are not sorry about it."

Contrary to what one might think, studies repeatedly show that snipers are well-adjusted, and score lower on tests for PTSD and higher on tests for job satisfaction than other military personnel. But at the same time these studies reveal that snipers are not immune from PTSD and that they could experience emotional problems in years to come.

The US Department of Veterans defines three sources of war-related trauma:

- being at risk of death or injury,
- seeing others hurt or killed,
- and having to kill or wound others.

Snipers suffer less of the first but more from the latter two.

When Soviet sniper Ilya Abishev fought in Afghanistan he believed that he was doing the right thing. His regrets came much later. "We believed we were defending the Afghan people," he says. "Now I am not proud, I am ashamed of my behavior."

One thing for certain is that only another sniper can understand what snipers experience.

Navy SEAL Sniper does an outstanding job of illustrating the life of a Navy SEAL sniper and helping the reader understand just what these brave warriors sacrifice both physically and mentally in defending our great nation.

Don D. Mann
author of the *New York Times* bestseller *Inside SEAL Team Six*

INTRODUCTION

Glen Doherty, my best friend and coauthor, former U.S. Navy SEAL, and perpetual adventure hound, was killed in action on September 11, 2012, in Benghazi, Libya. Glen, known to his friends as "Bub," was working as a Special Operations contractor in support of U.S. interests in the region. His unit came to the aid of Ambassador Stevens and his staff as their compound came under heavy attack. Glen's team fought against overwhelming odds to rescue the ambassador and over twenty State Department personnel. After transporting everyone to another compound, his team came under additional enemy rocket and mortar fire, and it was here that Glen gave his life in the defense of his fellow Americans. His operational excellence and aggressive fighting spirit ensured that the remaining team and Department of State personnel were evacuated safely. Glen's extraordinary heroism reflected the highest traditions of the United States Special Operations Forces.

Glen and I first met as new guys at SEAL Team 3 when we were both assigned to GOLF platoon. Like everyone who ever met Glen, I took an immediate liking to him. You couldn't help but like him; he just had that way about him.

Near the end of an eighteen-month platoon training cycle, Glen and I were called into the platoon office, both terrified that we'd done something wrong. In fact, it was quite the opposite. Our chief explained to us that the platoon was short on snipers and that we would both be given the opportunity to go through the Navy SEAL sniper training program, if we wanted to. We were both terrified—we knew how incredibly difficult it is to graduate from that program—and thrilled. Of course, we said yes.

⌃ **A great example of a twenty-first-century shooter/spotter pair on display; this is a long-distance and applied-technology configuration.**

The two of us were paired up together during the intensely stressful three-month-long Navy SEAL sniper program. We both agreed that it was also one of the most valuable training experiences a SEAL could have and that learning the sniper tradecraft makes you a better all-around operator. Our lifelong friendship was cemented during the long and painstaking hours we suffered through together during Naval Special Warfare sniper school.

After completing the course, we both deployed to the Middle East with GOLF platoon. The highlight of that deployment was when Glen and I were stationed as a sniper pair on board the U.S.S. *Cole* in October 2000, just hours after a suicide boat had blasted a hole in the hull and killed seventeen U.S. Navy sailors.

Our job was to set up on the bridge and ensure that perimeter security was not breached. This was accomplished by utilizing two MK11 RHIBs operated by the Special Boat Team and having trained snipers observe near and far with high-powered optics. Glen and I had a .50 cal each and several LAW rockets at our disposal. The *Cole* remained unharmed and was eventually returned to its home port.

We went our separate ways after that first deployment, but our close friendship continued both inside and outside the SEAL teams. Glen and I would often grab our mixed bag of friends and go on some crazy adventure. His friends were mine and vice versa, whether it was surfing Baja, flying (both of us were avid pilots), fishing for marlin in Cabo, skiing fresh powder, or writing this book together. It was always fun and never lacking in excitement.

I had just started my writing career and found myself having a book to deliver while raising a family and trying to hold down a day job. I needed help. When I asked Glen if he would help, he gave me a funny look and said, "What the hell, let's do it."

Glen and I were both frustrated with what was on the shelf in the way of sniper books. Most were written by non-snipers or self-proclaimed experts on sniping. These authors would opine at length about the art of sniping without ever having known firsthand what it's like to pull the trigger or call in close air support ordnance and drop it danger-close from a sniper hide.

Navy SEAL Sniper provides you with a plain-language look at the sniper craft from a Navy SEAL perspective—a sneak-peek look behind the curtain at what makes up a sniper's DNA. This book covers sniper history, current weapons, optics, gear, and missions and goes on to predict what the future of sniping will hold.

Our goal in writing *Navy SEAL Sniper* is to make you feel as if you're swapping stories with Glen and me over a shared campfire. After all, everyone was a friend to Glen, and he wouldn't want it any other way.

Glen is a true American hero and will be greatly missed by all. Still, I'm comforted knowing that a bit of him lives on forever in the pages of this book.

So, friend, pull up a chair, grab a cold drink, and join us as we take you into the world of the Navy SEAL sniper.

—Brandon Webb, October 2012

1

THE HISTORY OF SNIPERS

"Certainly there is no hunting like the hunting of man, and those who have hunted armed men long enough and liked it, never really care for anything else thereafter."
—Earnest Hemingway, from "On the Blue Water," *Esquire*, April 1936

Bring up the topic of snipers at your next dinner party and see where the conversation goes. It is a topic full of controversy and mystery, conjuring up a range of attitudes and images.

Hollywood has done its share to contribute to this fascination. An entire generation has grown up playing hyperrealistic video games that simulate combat sniper operations. Ask a middle-aged mother from New England and you might get a recap of an old *Cheers* episode where Sam Malone and his friends lure Dr. Crane out to the woods for an old-fashioned "snipe hunt," an old-school practical joke that leaves the mark deep in the woods with a sack, making clucking noises and hoping the elusive snipe will come jumping into the bag.

Darker connections to the term may remind some of the so-called Beltway Snipers, John Allen Muhammad and Lee Boyd Malvo, who terrorized the D.C. area in 2002, shooting innocent victims, and were eventually caught after one of the largest manhunts in modern history. Veterans of the armed forces may remember quiet men who kept mostly to themselves, carried scoped weapons, and often disappeared into the night, sometimes not returning for days on end.

The term *sniper* has a long history. In this chapter we will trace its roots back to the early days of marksmen, riflemen, sharpshooters, and hunters who well understood the power and potential that one well-aimed shot can have.

The projectile—an object propelled with great force—has played an important role in warfare since the beginning of time. Slings, spears, bows and arrows, crossbows, then later muskets and rifles, are all tools with specific applications, relevant for their particular time periods.

The bow begat the crossbow, which dominated the battlefield during the Middle Ages and was so effective that Pope Innocent II decreed the weapon "unfit for combat amongst Christians." This edict did not apply to combat with Muslims, however, so when King Richard the Lionheart took his army toward Jerusalem during the Third Crusade, he was well equipped with crossbowmen. His smaller force held off many attacks from Saladin's larger forces, thanks to their ability to maintain an accurate and rapid rate of fire.

Ironically, it was as King Richard was returning home from his failed crusade in 1199 that he was struck down by the very weapon his pope had sought to ban: Hit by a crossbow bolt from the ramparts of a castle he had under siege in Limousin, France, he later died of the wound. King Richard's death was the result of a medieval sniper conducting what would later become common practice on the battlefield: directly targeting leadership to affect command and control and demoralize those still alive.

≈ **King Richard was felled by a sniper's crossbow bolt.**

The Chinese began using gunpowder in the ninth century, but it didn't become important to European superpowers until much later, entering Europe in the thirteenth century, most likely through Arab trading routes. At first it was employed mostly in large cannons and siege engines, monstrous machines that hurled massive balls designed to knock down walls.

Over the next several hundred years technological improvements were made both to the powder and to the weapons in which it was used, developed from knowledge gained in the field by those who fought around the continent and throughout the world.

Early seventeenth-century wheel-lock pistol from England ≋

The flintlock musket eventually became a mainstay of the British and European armies, followed by the matchlock and then the wheel-lock mechanism, a development some believe was invented by Leonardo da Vinci.

≈ **Rifling in the barrel of a 105mm tank gun**

The difference between a rifle and a musket is that a musket has a smooth bore, whereas in a rifle, spiral grooves or "rifling" are cut into the inside of the barrel, imparting a spinning motion to the projectile, stabilizing its flight and allowing it to fly farther and with much greater accuracy. This gyroscopic stability of a spinning projectile had been known since the time of the bow and arrow; fins or vanes made of feathers or other materials, called *fletching*, were added onto arrows and crossbow bolts on a cant to impart spin and improve accuracy and range. It didn't take long for gunsmiths to apply this same knowledge to modern weapons.

In their early stages, the musket and rifle had separate and specific applications. The smoothbore musket remained the primary weapon for infantry because of its capacity to provide a trained soldier with a faster rate of fire. The ammunition for a musket was made to be slightly smaller than the diameter of the barrel, so the accurate range was most likely less than ninety meters. In a rifle, the ball had to be large enough to span the entire diameter of the barrel in order to engage the barrel's rifled grooves; because of this, the rifle took longer to load than the musket. In addition, the rifled grooves were prone to fouling from the gunpowder, so the barrel would often have to be cleaned between shots.

Fortunately for military tacticians of the time, the shooting style of the day was volley fire, that is, engaging large numbers of massed troops in a line shooting into other massed troops, one line firing while the other was reloading. In other words, it wasn't necessary to aim at any specific individual but simply at a mass of soldiers who would remain visible at close range even through all the smoke. After several rounds of volley fire, a bayonet charge could be expected.

Put yourself there for a moment:

You're shoulder to shoulder, wearing a wool jacket and hat, sweating like a pig under the hot sun. Your heavy musket is loaded, your hands are sweaty from nerves and the heat, and your mouth is dry. Cannons fire from your front and rear, and as you march forward, lines to your left and right are decimated by artillery fire. Smoke hovers over the field, fires burn in the distance, and men can be heard screaming in pain in all directions. Your ears are ringing, yet you maintain your cadence as you have been trained to do, taking direction from the officer to your flank.

Soon a blur of color appears through the haze, and a halt is called. You hear the command to "make ready!" and you cock your weapon and stand at the high port (weapon held aloft with both hands) as you hear a similar call shouted across the field.

At "Take aim!" you point your weapon toward the mass of troops.

"Fire!"

Loud claps of thunder seem to engulf the entire area as flashes from the gunpowder ignite in the muskets' pans, barrels erupt, and smoke fills the air. Men drop to your right and your left. Blood is everywhere, covering you as you seek to remember the natural rhythm of the reload. You measure the powder down the barrel, a different amount to the flash pan, place your ball and wadding in, ram it home, recock, and ready again.

To a modern soldier, especially a well-trained sniper, these tactics seem insane. How could anyone stand at such close range, receiving round after round of fire from the enemy, without seeking some sort of cover, or at the very least lowering his profile by getting down in the prone position?

The discipline of these troops, their dedication to duty and their fellow soldiers, and their pride and love of country truly must have been incredible. It is hard to imagine, if not impossible.

A modern military expression comes to mind: "If you're going to be stupid, you'd better be tough." Sounds about right.

While rifles were also common at the time, they were usually more expensive and were typically used in the countryside for hunting and by marksmen for sport and competition. With the discovery of the New World and the rapid colonization of new land by the European superpowers, the rifle soon proved invaluable for settling new and inhospitable country and forced old-world military tacticians to reevaluate their methods.

The Germans developed the leading rifle of the day, commonly called the Jaeger (hunter) rifle. Its short barrel fired a .50-caliber round or larger and was quite accurate for its day. Many of these weapons found their way to the colonies with German settlers, and in the heavily German area of Pennsylvania the Jaeger rifle evolved into what is now called the Pennsylvania long rifle, later called the Kentucky rifle when they were used by Kentucky sharpshooters against the British at the Battle of New Orleans during the War of 1812.

Life in the New World was hard. There was more land and more open space, which meant it took longer and more accurate shots to put meat on the table. The hunters and woodsmen of

⌃ **Volley fire**

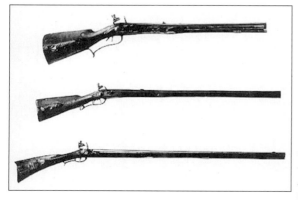

« **Top: a Jaeger rifle made by Andreas Staarman in late-seventeenth-century Berlin, with a 26-inch barrel and .75-caliber bore. Middle: a new-world hybrid owned by Pennsylvanian Edward Marshall in 1737; like the Jaeger, it has a patch box with a sliding wooden cover. Bottom: an early example of a fully evolved American long rifle; its 44.5-inch barrel made it more accurate than the heavy Jaeger, and its .44-caliber bore made it more efficient in its use of lead and powder. (*University of Virginia archives*)**

the day also had to adapt to and blend in with their environment, learn the art of camouflage, and develop stalking skills, patience, and the ability to make a successful kill with a single shot. All these skills would come into play in the wars to come and are still used today by the modern military sniper.

Prior to the American Revolution, several conflicts in the New World gave rise to evolving tactics and weapons as well as a new type of soldier—the ranger.

Rangers were woodsmen and hunters who operated in small units outside the safe confines of villages and forts. They engaged both hostile Native Americans and European forces that encroached on the new colonies. These men traveled light and fast, shedding weight and excess gear wherever possible, and adopted the fighting and raiding tactics of the Indians they fought against—guerrilla warfare in its earliest forms.

MAJOR ROBERT ROGERS,
Commander in Chief of the Indians in the Back-Settlements of AMERICA.

⌃ **Major Robert Rogers, founder of Rogers' Rangers**

« **June 21, 1759: Major Robert Rogers and his men scout the forest ahead of General Jeffrey Amherst's army on their way to capture Fort Carillon at Ticonderoga and other French posts on Lake Champlain (artist's rendering based on primary source descriptions). (*2010 Fort at No. 4, Educational Archive*)**

The skills inherent to the success of these units were passed on to the next generation, prior to the onset of the American Revolution. One early student who received the skills of these early Indian hunters was a man named Robert Rogers, who later became Major Rogers and led the famous Rogers' Rangers, formed in 1756 during the French and Indian War. His 1757 *Rogers' Rules of Ranging* is an illuminating read that vividly reveals the tactics of the conventional troops of the day. A fictionalized version of *Rogers' Rules* appeared in Kenneth Roberts's 1937 novel *Northwest Passage*, as explained to the narrator by Sergeant McNott, one of Rogers's men:

> All you need to know, I can tell you in ten minutes. I aint going to tell you twice, so listen to what I say, and remember it! Don't forget *nothing*! . . .
>
> Have your musket clean as a whistle, hatchet scoured, sixty rounds powder and ball, and be ready to march at a minute's warning.
>
> When you're on the march, act the way you would if you was sneaking up on a deer. Just remember the deer you're hunting now ain't going to run away if . . . he sees you first. You'll feel better afterwards if you see them first.
>
> You got to tell the truth about what you see and what you do. You can lie all you please when you tell other folks about the Rangers, but don't never lie to a Ranger or officer. There's an army depending on us for correct information.
>
> Always be careful! Don't never take a chance you don't have to.
>
> When we're on the march we march single file, far enough apart so one shot can't go through two men. If we strike swamps, or soft ground, we spread out abreast, so it's hard to track us. When we march, we keep moving till dark, so to give the enemy the least possible chance at us.

When we camp, half the party stays awake while the other half sleeps. If we take prisoners, we keep 'em separate till we've had time to examine 'em, so they can't cook up a story between 'em.

Don't never march home the same way. Take a different route so you won't be ambushed.

No matter whether we travel in big parties or little ones, each party has to keep a scout twenty meters ahead, twenty meters on each flank, and twenty meters in the rear, so the main body can't be surprised and wiped out.

Every night, when you're on the march, you'll be told where to meet in case we're surrounded [and] we have to scatter.

Don't sit down to eat without posting sentries. Don't sleep beyond dawn. Dawn's when French and Indians attack. Don't cross a river by a regular ford, because if anybody's laying out for you, that's where he'll lay. If you find out somebody's trailing you, make a circle, come back onto your own tracks, and ambush the folks that aim to ambush you. Don't stand up when the enemy's coming against you. Kneel down. Lie down. Hide behind a tree. Let him come till he's almost close enough to touch. Then let him have it, and jump out and finish him up with your hatchet.

That's simple enough, ain't it?

Many of the principles above are still very much in practice by SEAL snipers today.

If it ain't broke, don't fix it.

The American Revolution

During the American Revolution, British regulars were armed with what was called the Brown Bess, a smoothbore flintlock rifle, while the Hessian mercenaries carried short-barreled Jaeger rifles. The colonists used an assortment of weapons, including the Brown Bess and the very accurate Pennsylvania, or Kentucky,

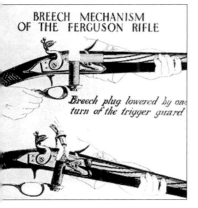

BREECH MECHANISM
OF THE FERGUSON RIFLE

Breech plug lowered by one turn of the trigger guard

≈ **Open-breech Ferguson rifle**

rifle. It was the first conflict in which sharpshooters were widely used, and it marked the beginning of the deliberate, systematic targeting of officers and other high-value targets.

Two parallel stories embody the outcome of the war and the role of snipers during that war—the famous "shot that was never taken" and a shot that *was*.

In the first, an inventor and marksman from Britain named Patrick Ferguson, then the captain of a unit of sharpshooters, was armed with his namesake, the breech-loading Ferguson rifle. Compared to the smoothbore musket, this rifle could maintain the same or greater rate of fire but without the common problems with fouling and reloading (it could be reloaded while on the move) that had plagued earlier attempts, and with amazing accuracy.

Prior to the Battle of Brandywine in September 1777, while in a hide site by a river, Ferguson and several of his soldiers witnessed an American officer and his French hussar cavalry companion ride within ninety meters, stop, and take into account the surrounding area.

Ferguson ordered his men to dispatch the two—easy shots with the Ferguson rifle—but then rescinded the order, considering it ungentlemanly to deliberately target fellow officers who were not directly bearing arms against them. In the pitched battle that followed, Ferguson was injured, and while recovering he learned that the target he had so chivalrously allowed to pass was none other than General George Washington.

≈ **One of Morgan's Sharpshooters armed with a Pennsylvania long rifle, later known as the Kentucky rifle (*Legacy Journal*)**

In direct counterpoint to Ferguson, the Colonials had Daniel Morgan, a talented tactician and marksman (rumored to be a distant cousin of Daniel Boone, the famous frontiersman and Indian fighter), who led specialized groups of shooters.

During the initial stages of the war, Morgan's Sharpshooters made a 950-kilometer forced march from Pennsylvania to Boston to engage the British.

Armed with Pennsylvania long rifles, Morgan's Eleventh Virginia Regiment of four hundred engaged in fierce fighting in October 1777 at the second battle of Saratoga. A British flanking unit led by General Simon Fraser was under heavy fire from Morgan's rifle corps, suffering devastating casualties and the disintegration of order and discipline.

As General Fraser galloped along in front of his lines to bolster his troops' morale and direct a counteroffensive, the American general Benedict Arnold gave Colonel Morgan the command to target Fraser and bring him down immediately.

No doubt several shooters were tracking and firing at Fraser as soon as the order was given, but it was a young sharpshooter named Timothy Murphy, perched in a tree and shooting at a range of approximately three hundred meters, who was credited with the kill.

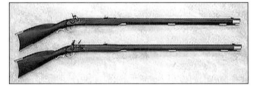

With their leadership gone, the British and Hessian troops folded and the Colonials won the day. It was a huge victory for the Americans, one that was celebrated even across the pond in Paris, France, at the time Britain's enemy.

Both Ferguson and Morgan found their way back into pivotal battles later on in the Revolutionary War. Ferguson met his end at the Battle of Kings Mountain—another huge victory for the Continental army—and Morgan, later promoted to general, won a resounding victory at the Battle of Cowpens against Colonel Tarleton, where legend has it he told his men, "Aim for the epaulettes."

Morgan continued promoting the use of sharpshooters throughout the remainder of his career and was instrumental in modifying long rifles to be fitted for bayonets, the lack of which

⌃ The Pennsylvania rifle was redubbed the Kentucky rifle after the Battle of New Orleans during the War of 1812, when a regiment from Kentucky used them to great effect against the British.

had been one major weakness the British had exploited during the War of Independence.

In 1929 FDR dedicated a statue in New York State to Timothy Murphy and during the ceremony he said the following which should resonate with soldiers everywhere:

> This country has been made by Timothy Murphys, the men in the ranks. Conditions here called for the qualities of the heart and head that Tim Murphy had in abundance. Our histories should tell us more of the men in the ranks, for it was to them, more than to the generals, that we were indebted for our military victories.

≈ **Baker rifle**

Sixtieth and Ninety-fifth riflemen shooting the Baker rifle (*Parks Canada*) ≈

The turn of the nineteenth century saw the rise of a military leader who brought with him a tactic that had not been used on a grand scale until his daring push across the European continent. Napoleon Bonaparte firmly believed in what would later be called blitzkrieg warfare: fast-moving, independent units rendering death and destruction, maneuvering quickly, operating independently, and wielding accurate rifle fire to its utmost potential.

To counter Napoleon, the British sent in the newly formed Ninety-fifth Rifles and the Sixtieth Regiment of Foot (later named the Rifle Brigade and the Kings Royal Rifle Corps), a brigade of the finest shots the crown could muster.

Using the new Baker rifle, these units went head-to-head with Napoleon. It's been said that the modern sniper expression "one shot, one kill" can be attributed to these marksmen.

These units had forsworn the old chivalrous code of battle and now specifically targeted officers, bugle man, drummers, and artillerymen. Many battles fought by these

⌃ The death of Admiral Nelson

new units produced casualty ratios as high as one officer for every five infantrymen killed, an astounding number considering the structure and tactics of standard military units. Once again the effectiveness of quality shooters proved their worth in the field.

A small victory for the French, despite Napoleon's eventual defeat, came at the hands of a sniper posted in the rigging of the French ship *Redoubtable*.

In October 1805 Admiral Lord Horatio Nelson was in command of H.M.S. *Victory* and engaged in a battle at close quarters with the French. It was a common tactic for the French, Spanish, and Portuguese during that era to send snipers aloft into the rigging to kill officers and take out gun crews. The British, allegedly due to Admiral Nelson's strategies, had not opted for those tactics and really hadn't needed to because Nelson had successfully crushed any naval opposition up until that point.

From a range of less than one hundred meters, Nelson was brought down by a French sniper, and despite the British winning the day, they lost their most seasoned naval commander.

© 2002 HowStuffWorks

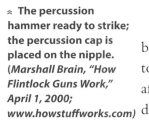

≳ The percussion hammer ready to strike; the percussion cap is placed on the nipple. (*Marshall Brain, "How Flintlock Guns Work," April 1, 2000; www.howstuffworks.com*)

Leading up to the Civil War, there were several technological advances that changed the way battles could be fought, along with vast improvements in the reliability and range of the modern rifle.

One of the first of these significant innovations was the development and implementation of the percussion cap, as opposed to the old flintlock system.

The expressions "flash in the pan" and "hang fire" both come from the antiquated and unreliable flintlock firing systems, where the flint ignited powder in the flash pan, sending a flame through the touchhole into the barrel. A "flash in the pan" occurred when a clogged touchhole caused only the primer charge to ignite: flash, but no bang. A "hang fire" resulted when the main charge at first failed to ignite but left a spark smoldering in the touchhole, which after a completely unpredictable number of seconds might suddenly explode, sending its bullet out the barrel. Not good.

In addition to solving these problems, the percussion cap also did away with the long follow-through required after a shot, that is, having to keep one's sight picture and alignment perfect for several seconds as successive charges finally sent the ball out of the barrel—easier said than done when holding a long, heavy weapon and taking effective fire from the enemy. With the percussion cap, misfires became extremely uncommon and wet weather became almost irrelevant to loading and firing.

Civil War–era percussion caps (*Vicksburg National Military Park, Carol M. Highsmith, photographer*) ≳

The percussion system used a small cap about the size of an eraser head, placed on top of a nipple with a pinhole that led directly into the barrel. When the hammer hit the cap, the flame was directed through the hole of the nipple straight into the barrel, where it ignited the main charge. Many of the flintlock rifles of the day were converted so that they could be used with this new ignition system.

The next major breakthrough changed ballistics forever. In 1847, Captain Claude Minié redesigned a round that had been originated in Britain but not adopted there. The minié ball is a conical-cylindrical self-expanding lead bullet that looks like a modern round. The original design had an expanding iron cup in its base that would be driven up into the lead and force it to engage the rifling grooves. Too often the iron cup would punch right through the lead, sending a mess of shrapnel out the end of the barrel.

≫ Human femur shot with a 510-grain lead minié ball fired from a .58-caliber Springfield Model 1862 rifle (*National Museum of Health and Medicine, Armed Forces Institute of Pathology, Washington, D.C.*)

An American named Burton perfected the design, although he never got name recognition, and the round that would forever be known as the minié ball would soon play a pivotal role in the bloodiest conflict the United States has ever seen.

NEW RIFLE-MUSKET BALL. Caliber-58

Weight, Ball 500 grains.
Weight, Powder 60 grains.

≫ Diagram of minié ball design

The results of this new round were phenomenal. Now the average soldier was capable of accurate fire out past 270 meters and volley fire past 900. Unfortunately, the tacticians leading troops during Europe's Crimean War and America's Civil War still employed the old tactic of massing troops shoulder to shoulder and exchanging fire at close range.

When I try to picture a modern SEAL platoon being ordered to form a line and shoot offhand (i.e., standing) at another platoon a few hundred meters away, I can imagine whoever delivered that order being the first one shot . . . and not necessarily by someone on the other side.

⌃ **The death of General John Sedgwick (*Library of Congress*)**

The American Civil War

Both the Union and Confederacy used sharpshooters effectively during the Civil War, and several generals from both sides suffered the same fate as General John Sedgwick, whose last words before being shot in the face by a Confederate sniper from a range of 720 meters were, "They wouldn't hit an elephant at this distance."

The Industrial Revolution was in progress, and the Springfield Armory in Massachusetts produced over one million Springfield rifles for the Union.

⌃ **General Sedgwick's corps awaiting orders, 1863**

⌄ Berdan's Sharpshooters featured in *Harper's Weekly*, October 5, 1861

⌃ Berdan's Sharpshooters working in an early shooter/spotter pair with security element

This reliable and effective rifle was truly a modern weapon for its day, with front and rear flip-up leaf sights and a maximum effective range out to 450 meters. The one real drawback to the Springfield was that it was still loaded from the muzzle, and the only easy way to muzzle-load quickly was standing up—not something one wants to do with Confederate marksmen in the field.

The Union had someone else pushing the envelope for battlefield deployment of trained shooters: a man named Hiram Berdan. Berdan was an engineer, an inventor, and a crack shot. He was also a wealthy man who had the ear of the president and secretary of war, and he convinced them to allow him to recruit a group of marksmen who became known as Berdan's Sharpshooters, consisting of two regiments formed by an active recruiting campaign throughout the North, with each candidate having to pass a rigorous marksmanship test.

The rifle that would eventually land in the hands of Berdan's troops was the famous Sharps rifle, a breech-loading rifle capable of incredible accuracy and reliability. It was often specially configured for Berdan's soldiers with a double trigger.

« The breech-loading Sharps rifle, weapon of choice of Berdan's Sharpshooters (*www.taylorsfirearms.com/products/bpSharps.tpl*)

« The .577-caliber Enfield P53, a favorite of the Confederate army (*www.cfspress.com/sharpshooters/arms.html*)

« The muzzle-loading Whitworth rifle, one of the best best long-range rifle of its era, sported a variety of sights, including a 4x telescopic scope. (*West Point Museum*)

Berdan himself was a bit of a boob who was not well respected as a combat commander, and he spent most of his time schmoozing in D.C., but his sharpshooters distinguished themselves throughout the war and produced plenty of impressive long-range kills of senior officers, artillerymen, NCOs, and regular troops.

The Confederacy was not without its own skilled shooters. At the beginning of the war the edge would most likely have gone to them because more of their troops came from rural communities, where being an accurate shot made the difference between eating and going hungry.

Early on in the war, the South also imported two excellent rifles, the British Pattern 53 Long Rifle-Musket, or Enfield P53, and the incredibly expensive and accurate Whitworth rifle. Confederate ordnance chief Josiah Gorgas called the muzzle-loading Enfield "the finest arm in the world." Sturdy, reliable, and quite accurate even at long range, it consistently outshot everything but the Whitworth and quickly became a favorite on both sides.

The Whitworth was a muzzle loader, with a hexagonal barrel that fired a special hexagonal bullet that sported a muzzle velocity of about 1,300 fps (feet per second). The base model reportedly had a purchase price of $600, with the top-end version, which included a telescopic scope and one thousand rounds of ammunition, going for $1,000—nearly $20,000 in 2009 currency. While not as good an all-around infantry weapon as the Enfield, this weapon was unparalleled for its time in long-range accuracy, reaching out as far as sixteen hundred meters, and was given only to the best of the best shooters from the South. The Whitfield is credited with multiple long-distance, high-value kills, including the famously

⌃ **Dead Confederate "sharpshooter" at Devil's Den; this photo was later found to be staged by the photographer.** (*Library of Congress*)

⚝ **Dead Confederate soldier in the trenches at Fort Mahone, Petersburg, Virginia** (*Library of Congress*)

During the First Boer War, the British took a pounding. ⌄

ill-timed (or, depending on your point of view, well-timed) shooting death of General Sedgwick.

Unfortunately for the South, they lacked the North's industrial complex, and importing excellent rifles like the Enfield and the Whitworth became harder and harder as the North's naval blockade grew more efficient and effective. As the war drew on, aside from the Springfields and Sharps they were able to acquire by picking through the dead, they had to bring the rest of their rifles from the farm or import them.

Despite significant success by both sides during the Civil War, the powers that be still did not embrace a philosophical change in tactics that would have more thoroughly used the special skills of these early breeds of snipers. Berdan's units were issued green and brown uniforms, and the Confederates were known to employ some early styles of ghillie suits, pinning leaves and brush to their clothes while stalking Federals. Still, this bloody war pushed technology to the limit, and a lot of the equipment and tactics of the modern sniper can be traced directly to the War Between the States.

The First Boer War

Excellent if somewhat lesser-known examples of the importance of sharpshooting in battle are the First and Second Boer Wars of the late 1800s.

The Boers were farmers, descendants of Dutch and German settlers living in two distinct areas of northeastern South Africa, the Transvaal and the

Orange Free State. Starting in December 1880, the Boers thoroughly embarrassed the British for the next eight months, fighting in a hit-and-run style with breech-loading rifles against the scarlet-clad British troops who, despite extensive combat experience, still had not adopted tactics capable of countering the Boers' accurate shooting and rapid movement.

Like the Confederates in the Civil War, the Boers were born with rifles in their hands and were experienced scouts and horsemen with a lifetime of field craft and local knowledge. During multiple battles, the Boer marksmen used a relatively new technique of "fire and movement" in which one group would keep the Brits' heads down by firing at them continuously while the other maneuvered to a more tactically advantageous position, and then the two groups would switch.

It didn't take long before the British withdrew, and the Boers of the Transvaal declared independence once again. The British were not done yet, though, and after licking their wounds, they returned with new weapons, new uniforms, new tactics, and a new resolve.

The Second Boer War

The second of the two Boer wars began in October 1899. For a decade, British citizens had been flooding into the Boer states to seek their fortunes in gold, which had recently been discovered there. The peace between the two groups had never been solid, and when conflict erupted once again, the Boers were ready. They had been purchasing large quantities of the bolt-action German Mauser rifle, which used recently perfected smokeless ammunition and was capable of holding ten rounds in an internal magazine. The British had adopted the Lee-Enfield, which was also a bolt-action rifle with an internal magazine but still used traditional black powder charges.

⤊ **Boers engage in some long-range shooting.**

The development of smokeless powder was a tremendous technological coup for gun and ammunition manufacturers, and it heralded the beginning of the true sniper era. Prior to this, even when firing from cover and concealment, a sharpshooter would give away his location by the large plume of smoke that erupted from his weapon after firing. Now there was only the muzzle flash, which can be very difficult to detect during daylight hours.

The Boers fought commando-style, organizing into small mobile units with democratically elected commanders and NCOs. They were totally self-sufficient, procuring their own food, water, and shelter as they harassed and killed the British.

In one particularly bloody battle at a hill called Spion Kop, also sometimes referred to as "the acre of murder," the British suffered some fifteen hundred casualties. In a single shallow trench on top of the hill, seventy-five British soldiers lay dead, each with a bullet hole in his head, victims of the rapid-firing Mauser and the Boer snipers.

Ultimately it came down to sheer numbers, and despite the Boers' ability to evade capture and continue their guerrilla campaign, a peace was finally negotiated in 1902, with shared control of South Africa going to the Boers and British.

World War I

When the Germans failed to quickly overrun France in 1914, what was then called the Great War quickly settled into a prehistoric war of attrition. Trench warfare was not a new concept, but both sides used new technologies or innovations such as mustard gas, hand grenades, new artillery shells, machine guns, and, of course, snipers to great effect in the slaughter.

⚈ **Battle-hardened Boers pose for a rare photograph. (*Hillegas,* With the Boer Forces)**

The term *sniper* is said to come from colonial India, where British officers hunted a small bird called the snipe, which is much like the North American woodcock. The snipe was quick and flew in erratic patterns, making it hard to spot. You had to be quite the marksman to shoot this bird, and those who developed a certain level of prowess in the sport were called snipers. Wherever the name came from, it was during World War I that it came into common usage.

For the first couple years of the Great War, German snipers dominated the western front, and British and French newspapers often related horrific stories of life in the trenches and how quickly the accurate fire of the German shooters could end a life.

⚈ **Germans fully geared up to repel a charge**

▴ **Gun crew from Regimental Headquarters Company, Twenty-third Infantry, firing a 37mm gun during an advance against German entrenched positions. U.S. Second Division, St. Michael region. (***RealWarPhotos.com***)**

The Germans were quick to appreciate the powerful effects snipers could have on lines that often were less than four hundred meters apart, and they equipped their best marksmen with some of the most technologically advanced rifles and optics in the world. Letters from British soldiers on the front were filled with stories of German snipers, telling tales of fellow soldiers taking the briefest of looks over the berm and falling backward, blood spilling out the back of their head.

Life in the trenches was hard enough, especially since the Germans had secured most of the high ground, with better drainage in their trenches, while the Allies' trenches were in the low ground, full of water and mud, vermin and death. Initially, the Germans also cut their trenches in a sawtooth pattern and used plenty of armor plating with loopholes to shoot through. The Allies' trenches were mostly straight, forcing the soldiers to stand up and set up their rifles on top of the meticulously level berm.

It took the British and French a while to realize the need to counter the German snipers, but they did eventually learn and things changed, although not quickly.

Several men in particular served as the catalysts for bringing the Allied forces to finally train and produce their own snipers. First and foremost was Major Hesketh Vernon Hesketh-Pritchard, who tried to get into the war at the age of thirty-eight but was denied, then managed to gain entry to the front anyway as a war correspondent.

Major Hesketh-Pritchard had an extensive background in big-game hunting and eventually

≈ **Flooded trenches along the western front (*National Library of Scotland*)**

« **Life in the trenches during World War I**

≈ **Effects of light and shade, from Major H. Hesketh-Pritchard's training course**
(*Hesketh-Pritchard,* Sniping in France)

pushed the need for Allied snipers up the resistant and overly bureaucratic chain of command. He was later commissioned in the British Army and returned to the trenches, where he and his unit went up and down the line, training soldiers in how to use their scoped rifles that had been indiscriminately handed out. What he found wasn't unexpected: No one was trained properly and most of the scopes were out of alignment or hadn't been zeroed.

It was a start, but he wasn't through yet. Eventually he got the high command to allow him to set up the very first sniper school, called the First Army SOS School (sniping, observation, and scouting). This became a seventeen-day course, covering much of the same curriculum as any modern-day sniper course would teach: rifle and scope setup and maintenance, ballistics, tactics, camouflage, observation skills, countersniper techniques, and, of course, marksmanship. Hesketh-Pritchard also established the shooter/spotter pair and believed (as I do) that it is actually much more difficult to be the spotter than the shooter. The initial course was

Brandon and I were shooting partners in sniper school and managed to post the high-est score for Navy personnel during the seven-week shooting phase of the course. We always managed higher scores with me on the trigger and him as the spotter. Brandon was gifted on the spotting scope, able to accurately call wind and adjust for environ-mentals no matter what the conditions. All I had to do was break a clean shot and have a good follow-through to give him an accurate call on where my shot broke.

Sniper school was unbelievably stressful, and I've often said that I would rather have gone through BUD/S (Basic Underwater Demolition/SEAL training) again than sniper school. In BUD/S I always felt if I gave it my all, things would work out. In sniper school, you could throw in everything including the kitchen sink—but have just one bad day and you were going home. And everything was graded.

We started with twenty-four SEALs in our sniper class; only twelve graduated. These were seasoned shooters, already wearing the Trident and specifically selected to attend the course because of their professionalism and marksmanship skills.

When people congratulated us later on successfully completing the course and on our relative standing as a shooting pair, I always gave the credit to Brandon. The man, as far as I am concerned, was unequaled as a spotter and fantastic on the trigger to boot.

Of course, he always credits me. I guess that's how a good shooter/spotter team works.

—Glen Doherty

≽ A contrast showing the drawbacks of a uniformed shooter versus the advantage of a well-camouflaged shooter (*Hesketh-Pritchard,* Sniping in France)

in Bethune, France, but was later moved to England, and was so successful that shortly thereafter a second army school of sniping was created.

Lovat Scouts using vegetation to their advantage »

Another British Army unit that contributed extensively to the newfound snipers' role in World War I was the Lovat Scouts, mostly formed from Scottish Highlanders. Many of these men had been gamekeepers, often called ghillies (from the Gaelic for "servant," as they would guide hunts for the rich), and had spent their days tracking animals, reading maps, observing, stalking deer, and practicing the art of camouflage. The modern-day ghillie suit traces its origins back to these men. They would don loose-fitting robes and use local vegetation woven into their suits to help them blend into their surroundings and effectively disappear.

A soldier in World War I uses a Springfield Armory modified rifle aimed through a periscope device. (*www.nps.gov*) ⩔

The ghillies perfected the art of camouflage, and their lessons were later taught at the first and second sniper schools, which would be attended by Americans after they joined the war in 1917.

In his *Sniper: A History of the U.S. Marksman*, author Martin Pegler illustrates the duties of a sniper as outlined in a pamphlet issued to American students:

1. To dominate the enemy snipers, thereby saving the lives of soldiers and causing casualties to the enemy.
2. To hit a small mark at a known range, but without the advantage of a sighting shot.
3. To keep the enemy's line under continual observation and to assist the Intelligence

⩔ **1903 Springfield with periscope attachment (*www.nps.gov*)**

of his unit by accurate and correct reports with map references.

4.　To build up and keep in repair his loopholes and major and minor sniping posts.

He also detailed this in simpler fashion as it was put out at the second sniper school:

1.　To shake the enemy's morale.
2.　To cause him casualties.
3.　To stop him working.
4.　To retaliate against his snipers.

Simple, to the point, and as relevant today as it was in World War I.

With the exception of the U.S. Marines, the American forces didn't really have much of an effective marksmanship training program in place, despite sending a few army officers and NCOs to the British sniper schools.

The U.S. Marine Corps had been heavily involved in competitive shooting since the turn of the century and because of this had continued to push the training envelope, a tradition that survives today. They distinguished themselves with their Springfield 1903s at the Battle of Belleau Wood, turning back the Germans with effective long-range rifle fire and suffering over one thousand casualties in the process.

In addition, many marine snipers were effective on the front with 1903s in combination with the Winchester A5 telescopic sight. The other sight commonly used by army snipers was the Warner & Swasey Model 1913, but the A5 was considered vastly superior and was still in use as late as World War II.

LA BRIGADE MARINE AMERICAINE AU BOIS DE BELLEAU
Dessin de GEORGES SCOTT.

≫ **Marine brigade at the Battle of Belleau Wood**

Rare photographic record of Gallipoli ≫

⌃ **An Aussie sniper uses a remote-controlled shooting system; his spotter helps key in on targets; calls wind and hits; and corrects for misses.**

Another front in World War I that saw the widespread use of snipers was Gallipoli, where Allied forces including the British, French, Australians, and New Zealanders tried to secure a foothold against the Ottoman Turks in an attempt to capture Istanbul and open a supply route to the beleaguered Russians on the eastern front.

The trench warfare on the front included the common usage of periscope sights and jury-rigged remote firing systems—scoped rifles on makeshift tripods with strings on their triggers, an Allied attempt to get accurate shots on Turkish shooters without exposing themselves to hostile fire.

Some famous shooters fought in Gallipoli, on both sides. The Australian Billy Sing went head-to-head with a famous Turkish sniper nicknamed Abdul the Terrible. The story goes that the sultan sent his foremost sniper after Sing, who had been decimating the Turkish lines (Sing was later credited with as many as 201 kills). After an intense battle over several days, as legend has it, the two located each other at the same moment, and Sing placed a shot right between Abdul's eyes a split second, before the Turk pulled the trigger. (If you haven't seen the movie *Gallipoli* starring a young Mel Gibson, it's worth a viewing—great ending.)

≈ **Turkish soldiers in a frontline trench at Anzac (*www.theage.com*)**

≈ **Relaxing in the trenches; suffering is more easily handled when shared.**

The Allies would eventually dominate the sniper battles on the western front, with the help of the fresh troops sent over from the United States. Unfortunately, after the war ended, the powers that be, who were reluctant to allow the sniper schools to exist in the first place, ended up dismantling them. There were still some who, despite the success of the Allied snipers, could not help thinking of it as "unsportsmanlike": truly a shame, as it would not be long before the snipers' services would be required again.

World War II

Most of the world's superpowers had all but dissolved their sniper programs when the German blitzkrieg stormed through Europe in 1939. Russia was the only country that claims to have maintained a full-blown sniper program in the interim, although the Germans still had plenty of shooters from World War I. The British had lowered their official snipers to eight per battalion, and among U.S. forces, only the Marine Corps maintained any sniper training at all, unofficially training just enough scout snipers to keep all knowledge from being lost.

There wasn't much place for snipers during the initial invasion through Western Europe, as the reliance on and application of speed, armor, and aircraft kept the battle away from the stagnated lines that were seen twenty-one years earlier in World War I. Snipers were mostly being used as a rear guard force, slowing down the rapidly advancing Germans. Most people remember Russia as an enemy of Hitler's Reich during World War II, but initially they had teamed with the Germans and invaded Poland together, decisively taking the country and dividing it between themselves.

Shortly thereafter, the Russians learned the hard way the deadly effect snipers and marksmen can have during what would later be known as the Winter War. Two

⌃ **Winter warfare brings about a whole new set of problems. Here Soviets prepare to counter an assault in Finland.**

Credited with over 500 confirmed kills, Häyhä likely had closer to 750. (*Za Rodinu*) ⌄

ZA RODINU

soldiers arose from this conflict posting numbers yet to be matched in the tallies of snipers throughout the history of war: Simo Häyhä and Suko Kolkka.

The Finnish people had a long history of marksmanship and hunting, and the inexperienced Soviet army of 1.5 million was easy picking for the camouflaged shooters, who operated in familiar terrain, often on skis, in a motivated defense of their homeland.

Häyhä is credited with over five hundred kills with his Moisin–Nagant Model 28 rifle (using open sights), and another two hundred using other weapons. He was shot through the face in March 1940 but managed to survive and recover from his wounds, receiving a field promotion from corporal to lieutenant, the highest jump in rank in Finnish military history. Later in life he was asked how he became such a good shot. His reply—you have to love this: "Practice."

≈ **Simo Häyhä notched an incredible tally of kills, mostly using iron sights.**

≫ **The relatively green Soviet troops suffered badly against the Finnish defenders.**

Suko Kolkka's tallies are also incredibly impressive and he is mentioned in many books on snipers, but his actual existence is in question. There are no records of him in any Finnish military personnel files, and it is now considered possible that he is strictly legend.

With or without Kolkka, the Finns decimated the Soviets, bringing their casualty numbers to the one-million mark (while suffering only twenty-five thousand Finnish casualties and forcing the Soviets to the negotiating table. In the end, a treaty was signed, with the Finns giving up some land and resources to the Russians.

The following comment from one Soviet general after the treaty was signed has been quoted in many books on snipers, but it's worth repeating again here: "We gained 22,000 square miles of territory . . . just enough to bury our dead."

Once the war was in full swing, all the major players got back in the game and started training snipers again. The British opened

≈ An example of great winter camo during World War II

Barry Pepper playing Private Jackson in *Saving Private Ryan* sighting in with the 1903A4. Okay, it's Hollywood—but he was a great character, and shot lefty to boot. ≈

a school in Bisley, with the help of the famous Lovat Scouts, and provided marksmen with a three-week crash course in sniping: camouflage, observation skills, advanced marksmanship, fieldcraft, and scouting.

The British primarily depended on the Lee-Enfield No. 4 with a No. 32 scope, which proved to be a great rifle throughout the war and beyond.

In the United States, an advanced instructor's rifle course was established at Camp Perry in Ohio, and a sniper school was formed in late 1942 at Fort Bragg, North Carolina, in addition to programs at Camp Lejeune, Camp Pendleton, and Greens Farm in San Diego.

Unfortunately, most of the U.S. Army schools focused primarily on marksmanship and neglected to emphasize many of the other important skills essential to any competent sniper. The other important skills were learned in the field, from other snipers, or in forward training environments like the sniper schools set up in North Africa while the Allies battled Rommel prior to the invasion of Sicily.

≋ **German sniper in World War II**

The U.S. Marine Corps fortunately recognized the importance of teaching the "scout" aspect of the title given to scout snipers. In addition to marksmanship, they focused on camouflage, reconnaissance, map-reading, directing artillery, and observation.

Throughout the war the primary sniper rifle of the Americans was some version of the 1903 Springfield, mostly the M1903A4. With infantry carrying the fantastic M1 Garand, it's a wonder that weapon could not be converted for snipers, but unfortunately a good scope mounting system wouldn't be sorted out until late 1944, so the 1903 became the rifle of choice, accompanied by the 2.5x Weaver 330C scope (mostly used by U.S. Army snipers) or the Unertl 8x scope used by the U.S. Marine Corps.

⌃ **German sniper in World War II with the Mauser 98k**

Under the orders of Heinrich Himmler, the Germans also began a sniper program in Zossen. In Pat Farey and Mark Spicer's *Sniping: An Illustrated History*, the authors detail the German "Snipers' Code" as follows:

1. Fight fanatically.
2. Shoot calm and contemplated; fast shots lead nowhere; concentrate on the hit.
3. Your greatest opponent is the enemy sniper; outsmart him.
4. Always only fire one shot from your position; if not, you will be discovered.
5. The entrenching tool prolongs your life.
6. Practice in distance judging.
7. Become a master in camouflage and terrain usage.

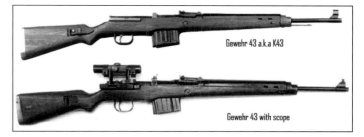

Gewehr 43 a.k.a K43

Gewehr 43 with scope

⌃ **German Gewehr 43 rifles**

8. Practice your shooting skills constantly, behind the front and in the homeland.
9. Never let go of your sniper rifle.
10. Survival is ten times camouflage and one time firing.

The Germans were primarily using the Mauser Kar 98k bolt-action rifle, fitted with a variety of different sights including the Zf41, which reminds me of a lot of modern quick-acquisition target sights such as the ACOG, Aimpoint, or EOTech. The Germans also experimented with one of the first semiautomatic sniper rifles, the gas-operated Gew43, which was tactically superior in close combat but not as accurate at range.

In 1941, the teams that once played nice turned on each other, and Germany invaded Russia.

Now Russian snipers were used to great effect as a rear guard behind the retreating Russian army, taking out German officers, NCOs, and artillerymen as well as lines of communication. The Russians used primarily the Moisin-Nagant 1891/1930 rifle with scopes made by Zeiss—primarily a 4x scope at first, though they later mass-produced a 3.5x telescopic sight that was well received.

The record-breaking sniping numbers of the Russian military are extraordinary, with a tremendous list of men and women snipers, each tallying over three hundred confirmed kills. Knowing what we do about the Soviet propaganda machine, it is likely that these numbers were exaggerated, but even if you cut some of their top snipers' numbers in quarters, they are still quite impressive. The Soviets relied on the "cult" of snipers as part of their propaganda program and as a morale

⌄ **The Russian Valkyrie**

booster for their beleaguered troops, and they idolized these men and women in the press.

The Soviets were also the first to include women in their snipers ranks, some of whom became legendary. One female sniper named Lyudmila Pavlichenko (whom the Germans called "the Russian Valkyrie") fought in the defense of Odessa and the Crimea and was invited by Eleanor Roosevelt to the United States to speak to people about the war on the eastern front.

In Andy Dougan's *Through the Crosshairs: A History of Snipers*, he mentions a speech Pavlichenko gave in Chicago, during which she said, "I am twenty-two years old, and I have already destroyed 309 enemy soldiers who have invaded my country. I hope you will not hide behind my back for too long."

The Soviet sniper we have all heard of, read about, and seen in the Hollywood movie *Enemy at the Gates*

⋆ **Lyudmila Pavlichenko during happier days**

⋆ **Female Russian sniper after being on the losing end of a battle with a German sniper (*Za Rodinu*)**

⌃ **Zaitsev shows his rifle to a Russian general.**

(which would have been a great film if they could have left the love story out of it—just one guy's opinion) was Vasily Grigor-yevich Zaitsev, who was credited with over two hundred kills in

Vasily Zaitsev teaches new snipers with a Moisin-Nagant and 4x power PE scope. (*Beevor, Antony:* The Fateful Siege) »

the defense of Stalingrad alone and was later awarded the decoration of Hero of the Soviet Union.

The shepherd from the Urals improvised an academy within the besieged Stalingrad to help train other snipers, but he is most famous for a duel that allegedly took place between himself and a German officer. If you've seen the movie, then you know how that duel went, but whether or not the Germans sent in their top sniper to deal with Zaitsev, his diaries do detail a multiple-day stalk in which he and two of his shooter/spotters went head-to-head with a German sniper and ended up killing him as he hid beneath a sheet of metal in a contested area.

The bloody war on the eastern front, what the Russians call the Great Patriotic War, was the deadliest battleground in modern history, with an estimated twenty-five million Russian soldiers and civilians dead and three million to five million Germans killed. Thousands of these deaths were caused by single shots from a sniper's rifle.

⌃ **Running and gunning in Stalingrad**

The Germans also had prolific killers, and once the tide turned and the Russians were on the attack, German snipers took on the duties of the rear guard. In *Sniper on the Eastern Front: The Memoirs of Sepp Allerberger*, famed German sniper Allerberger tells author Albrecht Wacker about some of his tactics for stopping an advance:

> I would bide my time until the next four waves were on their way towards our lines, then open up rapid fire into the two rear waves, aiming for the stomach. The unexpected casualties at the rear, and the terrible cries of the most seriously wounded, tended to collapse the rear lines and so disconcert the two leading ranks that the whole attack would begin to falter. At this point I could now concentrate on the two leading waves, dispatching those Soviets closer than fifty meters with a shot to the heart or the head. Enemy soldiers who turned and ran transformed into men screaming with pain with a shot to the kidneys. At this, an attack would frequently disintegrate altogether.

⌃ **Infantry duke it out in the hedgerows of France.** (*Medical Service in the European Theater of Operations*)

D-Day

The battle for Western Europe took on a new face with the launching of Operation Overlord, what most of us remember as D-Day. The German snipers wreaked havoc on the Allied troops, who were fighting a new kind of war in terrain the Normans call bocage.

Fighting through the hedgerow country of France was hell for the Allies, as German snipers decimated the ranks of officers, NCOs, and medics to the point where officers began to disguise themselves, carrying the weapons of the enlisted troops, hiding their rank insignia, and keeping their binoculars concealed. On the beaches the corpsmen were easily targeted, distinguishable by the white band with a red cross on their arms, and they were often gunned down while dragging wounded soldiers to cover.

The thick foliage that bordered the roads and farms was perfect for cloaking snipers, and the German command took full advantage. A single sniper could hold up entire columns for hours or more, and new countersniper tactics had to be brought to bear to keep the front moving toward Germany.

The Germans would send out one- to three-man teams, with enough food and water for several days. They operated ahead of or behind the lines at their own discretion; often when supplies ran out they simply surrendered. One can imagine that many a kangaroo court dispensed swift battlefield justice to captured snipers. Snipers were widely reviled and often didn't survive to see the inside of a POW camp.

Pulitzer Prize–winning journalist Ernie Pyle, one of the great war correspondents of World War II, had

≋ **American soldier killed by German sniper, Leipzig, April 18, 1945 (*Robert Capa, gelatin silver print: Cornell Capa International Center of Photography*)**

Soldiers man a machine gun at the base of a hedgerow. (*82nd Airborne Museum*) ≋

≈ A road between two hedgerows near Normandy; it would be pretty hard to pick out a camouflaged shooter in that. (Can you see one?) (*www .normandybattlefields .com*)

this to say about sniping in the hedgerow country of France, written in a report filed a few days after D-Day:

> Sniping, as far as I know, is recognized as a legitimate means of warfare. And yet there is something sneaking about it that outrages the American sense of fairness.
>
> I had never sensed this before we landed in France and began pushing the Germans back. We had had snipers before—in Bezerte and Cassino and lots of other places, but always on a small scale. There in Normandy the Germans went in for sniping in a wholesale manner. There were snipers everywhere; in trees, in buildings, in piles of wreckage, in the grass. But mainly they were in the high, bushy hedgerows that form the fences of all the Norman fields and line every roadside and lane.

It was perfect sniping country. A man could hide himself in the thick fence-row shrubbery with several days' rations and it was like hunting a needle in a haystack to find him. Every mile we advanced there were dozens of snipers left behind us. They picked off our soldiers one by one as they walked down the roads or across the fields. It wasn't safe to move into a new bivouac area until the snipers had been cleaned out. The first bivouac I moved into had shots ringing through it for a full day before all the hidden gunmen were rounded up. It gave me the same spooky feeling that I got on moving into a place I suspected of being sown with mines.

In past campaigns our soldiers would talk about the occasional snipers with contempt and disgust. But in France sniping became more important, and taking precautions against it was something we had to learn and learn fast. One officer friend of mine said, "Individual soldiers have become

≳ **German sniper in Normandy**

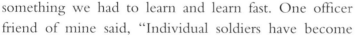

« **Ernie Pyle was killed in the Pacific theater on the small island of Ie-shima off Okinawa; shown here, he is being buried along with other soldiers killed during the assault.**

sniper-wise before, but now we're sniper conscious as whole units."

Snipers killed as many Americans as they could, and when their food and ammunition ran out they surrendered. Our men felt that wasn't quite ethical. The average American soldier had little feeling against the average German soldier who fought an open fight and lost. But his feelings about the sneaking snipers can't be put into print. He was learning how to kill the snipers before the time came for them to surrender.

The Pacific Theater

Both the U.S. Army and marine Corps saw intense combat in the Pacific, with the marines doing most of the island hopping and the army seeing its action in the Philippines and on the Southeast Asian mainland.

≋ **Beach assault in Saipan (Life's Picture History of World War II)**

↟ **Marine snipers in Okinawa scan for Japanese soldiers with their 1903s and Unertl scopes. Shortly after this photo was taken, one of the shooters allegedly dropped a Japanese soldier at nine hundred meters.**

On many of the islands taken, it was difficult to employ the sniper with traditional equipment because of the thick jungle environment. Other islands, such as Okinawa, Saipan, and Iwo Jima, were much more open, and there are plenty of recorded stories of U.S. Marine snipers taking out the enemy at ranges greater than nine hundred meters.

The Japanese snipers were using primarily the Ariska Type 97 sniper rifle, which shot a 6.5mm round and used a 2.5x telescopic sight. Later in the war, a shorter-barreled rifle was released, the Ariska Type 99, which transitioned to a larger 7.7mm round and had a more powerful scope on it as well.

≽ **Marines check out dead Japanese snipers off the island of Leyte in the Philippines.**

The Japanese snipers followed the Bushido code and were honor-bound to accept death before surrendering in the name of their emperor and homeland. In the tight jungle warfare, they were often found high in palm trees, well camouflaged and waiting for the troops below to pass before targeting officers and NCOs from behind to create more havoc.

Iron sights and machine guns were particularly useful in the close quarters of jungle warfare. U.S. forces would handle the Japanese snipers more often than not by spraying the treetops with the Browning automatic rifle (BAR) or Thompson machine gun. If they were taking fire from farther in front, with a clear field of fire, then they could employ countersnipers, often taking out enemy machine-gun nests with accurate fire from over a kilometer out.

Later, toward the end of the war, it was not uncommon for the Allies to call in heavy artillery and completely level an entire area just to clear out one sniper. Necessity is the mother of invention; nothing was out of the realm of possibility. Dogs

⌃ **U.S. troops check out a Japanese snipers' nest on Papua New Guinea.** (*Rickard, "Japanese Snipers' Nest," New Guinea, November 17, 2008*)

⌃ **Marines advance with the Thompson machine gun and the BAR, both great weapons used to deadly effect countering Japanese snipers. (*National Archives*)**

were used, grenades, satchel charges, howitzers, antitank recoilless rifles—you name it. But when in doubt, blast them out.

The Japanese may not have been the best marksmen, but they were renowned for their skills in camouflage and jungle survival. Farey and Spicer's *Sniping: An Illustrated History* quotes a 1943 *Time* article:

> Marine and Army men returning from the South Pacific almost unanimously hold that, man for man, the Jap soldier is inferior in fighting qualities to the American. But in all the things to do with hiding, stealth, and trickery, they give the Japs plenty of angry credit.

They dig deep, stand-up foxholes, which are safe except under direct artillery fire (and which are better than U.S. slit trenches). On the defensive, they dig themselves dugouts protected by palm trunks, and then they crawl in and resist until some explosive or a human terrier kills them.

The myth of the Japanese sniper is exploded by returning officers. They say that Japanese snipers are an annoyance, little more. They hide excellently but their aim is poor. Sniping serves, however, to frighten men who will not deliberately ignore it. . . .

But the greatest handicap of the Japanese is their lack of imagination. They carry out orders to the letter and, if necessary, to death. But when things go wrong, they cannot adapt their tactics. If Jap attackers meet resistance, they advance anyhow—which accounts for the terrible slaughter to which Japanese troops submit themselves.

The "terrible slaughter" of the Japanese had yet to reach its zenith, and it would not be the sniper that would bring an end to the war in the Pacific.

On August 6, 1945, a B-29 named the *Enola Gay* dropped warfare's first atomic bomb, dubbed Little Boy, on Hiroshima. Three days later, a second bomb leveled Nagasaki, and shortly thereafter World War II came to an end.

Korea

The peace was tenuous in the years following World War II as the world witnessed the arrival of two superpowers that would spend the next forty years duking it out over the hearts, minds, and bodies of the world's population. Lines were drawn and the Western Allies geared up to stem the tide of Communism.

In the new Atomic Age, little thought was given (once again) to the relevance of the lone shooter as the Allies thumbed their

≈ **Moison-Nagant sniper rifle with PU scope**

noses at history and allowed their sniper programs to go the way of the dodo bird. Not so the Communists. The Russians had witnessed and highly valued the destruction their shooters inflicted on the German ranks, and they passed on their tactics to other Communist states.

In June 1950, the fragile peace that existed between Communist North Korea and Democratic South Korea dissipated as the North pushed below the 38th parallel, sparking a bloody, three-year war.

Initially, the Chicom (Chinese and North Korean) soldiers wreaked havoc on the Allies with accurate sniper fire from primarily World War II–era Moisin-Nagant sniper rifles with PU or PE scopes.

As usual, despite pressure from the actual combatants, a senior officer had to be personally affected before anything got done. One story from the war tells of the start of the U.S. in-house sniper program in Korea.

In his book *The Sniper at War*, Michael Haskew retells a classic story from military author Adrian Gilbert. The commander of the Third Battalion, First Marines allegedly was trying to survey the field one morning when, in Haskew's words:

He placed his binoculars in the bunker opening and gazed out. Ping! A sniper bullet smashed the binoculars to the deck while blood welled up in the crease in his hand.

The battalion commander, fortunately, was only scratched, but he reflected that it was a helluva situation when the CO could not even

Chinese sniper Zhang Taofang allegedly had 214 kills in just over a month using an old Moisin-Nagant with iron sights. ≈

⚠ **USMC sniper pair working the field in Korea (*Department of Defense Visualization Center*)**

take a look at the ground he was defending without getting shot at. Right then and there he decided that something had to be done about the enemy sniper.

The colonel learned that in the supply section there were an adequate number of rifles and telescopic sights. He next sent for an experienced gunnery sergeant who had spent time firing with rifle teams. He told the gunny what he wanted.

The gunny visited each company to pick sniper candidates. He wanted riflemen who possessed the characteristics of good infantrymen. But above all, he stressed the need for

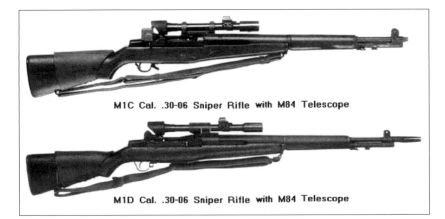

M1C Cal. .30-06 Sniper Rifle with M84 Telescope

M1D Cal. .30-06 Sniper Rifle with M84 Telescope

patience. This trait is absolutely essential, for a sniper must remain still and alert for long hours, waiting for the enemy to show himself.

Soon the range was ready, and the gunny began an intensive three-week course on sniping . . . after training, the Marine teams were sent out to the various outposts. To spur them on, a case of cold beer was awarded to the men of each outpost that got twelve kills within a week. All hands turned to helping the rifle experts in spotting enemy snipers.

Only a week after the sniper teams went into action, the division commander came to test their efficiency. The two-star general strode the entire length of I company's main line of resistance armed with nothing but his walking stick. . . .

The Americans had to dig into the storage units to get out-fitted for Korea and went back to using the M1C Garand semiautomatic or the Springfield 1903 bolt action.

In close combat, it was invaluable to have a quick follow-on shot, but over longer ranges, the 1903 was still preferred. The M1 suffered from two major drawbacks. First and fore-most, it was still loaded by a metal clip that would eject with a loud "clang" when the last round was fired—not so good when you are trying to conceal your location. Also, the enemy would know you were reloading. Secondly, the optics still had to be mounted off to the side, which didn't make for ideal body position and eye relief.

Modifications were becoming available though; add-ons in the form of leather pads for the stock for more appropriate cheek welds and flash suppressors were becoming readily available to full-time shooters.

In-country armorers also undertook several interesting experiments. Modifications were made to the Browning automatic rifle, such as mounting a Unertl 8x scope and firing the weapon in single-shot mode off a bipod or tripod, which proved accurate to eighteen hundred meters. Other talented inventors played with captured Soviet antitank rifle actions and Browning barrels to create various .50-caliber long-range sniper weapons.

In *One Shot—One Kill*, Charles Sasser and Craig Roberts tell another classic story of sniping on the front lines during the Korean War. U.S. Army Corporal Chet Hamilton was watching the assault of a steep hill (near the famous Pork Chop Hill) and was in position to see his boys take a beating from the Chinese Communist troops, known at the time as Chicoms:

> I felt helpless watching from the sandbagged trenches . . . until I noticed something. It was only about 400 yards across the valley from the Chinese lines. My position put me on almost the same level with the Chink defenders on the other hill.
>
> In order for the Chicoms to see our troops and fire at them down through their wire as the GIs charged up the hill, they had to lean up and out over their trenches, exposing wide patches of their quilted hides. (Chicoms often wore quilted uniforms for protection during the ridiculously cold winter months. The Allies weren't so fortunate.)

That was all I needed. It had become a clear morning in spite of the smoke and dust boiling above the Chinese hill. The four-power magnification of my scope made the chinks leap right into my face.

All I had to do was go down the trench line, settle the post-and-horizontal-line reticle on one target right after the other, and squeeze the trigger.

It was a lot like going to a carnival and shooting those little toy crows off the fence. Bap! The crow disappeared and you moved over to the next crows. By the time you got to the end of the fence, you came back to the beginning and the crows were all lined up again ready for you to start over. I don't know who the Chinese first sergeant was over there, but he kept throwing up another crow for me every minute or two. And I kept knocking them off the fence. The fight for the hill lasted about two hours. The other guys . . . came to watch, point out targets, and cheer when I zapped one.

The shortsightedness of U.S. military commanders continued after the signing of a peace treaty in 1953, and the art of the seasoned military sniper was once more shelved in the practicality closet and replaced with the idea of global thermonuclear war. Some advocates, notably Warrant Charles Terry and Second Lieutenant Jim Land who were shooting on the U.S. Marine Corps team, fought to keep the hard-earned skill sets alive and well in the U.S. military. Those skills would be needed again soon.

Vietnam

Leave it to the careerists and politicians to ruin a perfectly good thing. I realize I'm probably beating a dead horse here, but in every major conflict in which the United States was involved, the sharpshooters and snipers proved their value time and time again, only to have politics and weak leadership inevitably intervene and

▲ **Vietnam-era XM21 sniper system**

squash whatever programs were in place, pushing the true battlefield commanders behind desks and letting the rest of the talent go where they may. Vietnam wasn't much different.

The U.S. Marines were the first conventional troops to enter Vietnam in March 1965, and they quickly came under accurate and constant fire from seasoned Vietcong (VC) guerrilla fighters and North Vietnamese army (NVA) regulars who had been fighting for independence for over a decade.

This was jungle warfare all over again, and throughout the course of the war there were very few large-scale engagements with two armies toe-to-toe in a pitched battle. It was guerrilla fighting, hit and run, ambushes, booby traps, and highly trained snipers.

Captured instruction manuals give us a look into what was a well-run and intensive North Vietnamese sniper course, most likely aided by Soviet advisers. In addition to long-range shooting they also taught basic armorer skills so that shooters could repair

Vietnam snipers; note the XM21 suppressors and optics, cutting-edge for their time. ⤋

their own weapons in the field. These snipers were also taught camouflage, observation, and reconnaissance and would take these skills south to teach to their guerrilla counterparts, the VC.

A great example of the effectiveness of one well-trained sniper is given in Adrian Gilbert's book *Sniper*:

One such instance was observed by a Marine sniper, Joseph T. Ward, who described how a communist sniper had pinned down a neighboring company of Marines.

Although an air strike had been called, low clouds prevented the three-strong flight of F-4 phantoms from attacking for over two hours. During this time, whenever a Marine moved he was picked off by the sniper. Even after the first strike of napalm, the sniper continued his work, wounding another man before departing the scene as the second strike went in.

⋩ **Painting of a sniper in Vietnam (*National Museum of the Marine Corps*)**

Obliged to admire the skill of the persistent sniper, Ward recalled: "While we cleaned our rifles, I thought what a day's worth one of Uncle Ho's best had given. One enemy sniper had killed three grunts, wounded four, tied up two companies and six fighter-bombers a good part of the day, and we hadn't seen any sign of him."

As there was no official scout sniper program in place, other than a small program run by Captain Jim Land in Hawaii in the early 1960s, a program was quickly set up in country, recruiting veterans of the U.S. Marine shooting team as instructors. With pressure from above, Captain Bob Russell and later Captain Land each set up marine sniper schools in country, both within the vicinity of Da Nang, bulldozing nine-hundred-meter ranges into the bush and creating their own curriculum from lessons they learned by personally taking out VC in the field.

Land was under strict orders from the commander of the First Marine Division: "I want you to organize a sniper unit within the First Division. Captain Russell in the Third Division started training snipers last year. I want mine to be the best in the Marine Corps. I want them killing VC and I don't care how they do it, even if you have to go out there and do it yourself."

↟ **Signed photo of the legendary Carlos Hathcock**

Hathcock was credited with hitting an NVA at overeighteen hundred meters with a .50-cal mounted with a telescopic scope and set to fire on single shot. ⩗

One of Land's first students was a bored former MP who would later become a sniping legend and true hero: Carlos Hathcock, also known as the White Feather, or Long Tra'ng in Vietnamese, because he wore a white feather in his hat.

Already a decorated competition shooter, Hathcock was a natural for the job and produced some of the most impressive stalks, kills, and survival stories to come out of the Vietnam War. Some of his shooting highlights include 93 confirmed kills and a shot with a Browning .50-caliber modified with a scope that dropped a VC at two thousand meters. There were snipers with more confirmed kills (U.S. Marine Chuck Mahwinney: 98, U.S. Army sniper Adelbert Waldren: 109), but none were as well-known.

Hathcock's most heroic act didn't even involve a rifle. While traveling in a half-track with some other marines, the convoy was shelled and his vehicle burst into flames. Despite being on fire, he helped rescue all the other marines in the truck, before barely escaping himself, with third-degree burns covering 40 percent of his body. It took over a year for him to recover.

This heroic act earned Hathcock the Silver Star and also effectively ended his active sniping career. However, his knowledge and experience were not allowed to go to waste (for once!), and he later became an instructor at the U.S. Marine Corps sniper school in Quantico, Virginia, which opened in 1977.

The Marine Corps' push for new snipers also eventually brought about a new weapon, the Remington 700, soon to be known as the M40 sniper rifle, with the Redfield Accurange 3–9x scope replacing the old Winchester Model 70.

(The Remington is a fantastic weapon and still in use today.)

The U.S. Army, typically bringing up the rear (kidding, Army brothers, kidding), didn't fully adopt a sniper program until June 1968 when they brought members of the AMTU (Army Marksmanship Training Unit) from Fort Benning, Georgia, to Vietnam. The AMTU guys showed up with a few leftover sniper rifles, some old Garand M1C and M1Ds, along with the classic Springfield M1903 A4 paired with the incredibly average 2.2x M84 telescopic sight, and proceeded to establish a difficult and intensive eighteen-day training program.

Back in the States, meanwhile, the push was on to find a new weapon. After extensive testing at home, the army settled on what would become the XM21, a match-grade M14 fitted with the Redfield 3–9x power scope with some cool new technology called ART (Automatic Ranging System). This system essentially eliminated the need to range the target inside of six hundred meters by automatically adjusting for the range, so long as the shooter put the target's body between two lines in appropriate fashion using a special cam mechanism.

Other technology, including new night-vision optics and suppressors for the rifles, proved to be extremely effective in the field; this is where the United States truly began to "own the night," as the common sniper's expression goes.

It had taken a while, but the army finally realized what the marines had so aptly summed up on a sign outside their sniper school in Quantico: "The average rounds expended per kill in Vietnam with the M-16 was 50,000. Snipers averaged 1.3 rounds. The cost difference was $2,300 vs. 27 cents." Despite producing highly qualified snipers in country, however, the big Green Machine could

≈ Sniper Adelbert Waldron, Vietnam, 1969 (*www.olive-drab.com*)

Portrait of PFC Chuck Mahwinney (*Michal W. Wooten © 2003 Military Art Gallery*) ≈

not figure out how to employ them at first. In Martin Pegler's *Sniper: A History of the U.S. Marksman*, a U.S. Army private sums up the situation perfectly:

> Although I trained as a sniper I was just another dog-faced grunt in the squad, pulling patrols and guard. I didn't even get issued a sniper rifle, only the M16. When we got hit by a VC sniper the lieutenant started yelling for me, but with the M16 I couldn't do anything.
>
> I went to the captain and told him sending me on a sniper course then using me as a grunt was stupid. He agreed and I got an immediate issue of an XM-21, which really pissed the lieutenant off. He kept putting me on point and it was some weeks before they started pairing us up and really using us for what we'd been trained for.

≋ **Lance Corporal Dalton Gunderson, USMC, demonstrates a modified seated shooting position.**

Once they were properly employed, the army snipers were incredibly effective, as they proved in the first six months of 1969, when they racked up some 1,139 confirmed kills. Cost-effective, deliberate, low-maintenance, and deadly: That was the U.S. sniper in Vietnam.

To the Present

Since Vietnam, the United States has been involved in quite a few conflicts while policing the world: Beirut, Grenada, Panama, Somalia, Bosnia, Afghanistan, and Iraq, to name a few. The U.S. sniper played a part in all of these, and great stories have come out of each conflict.

At the same time, the professional sniper-soldier has a new and ever-growing problem: politically correct war fighting. Since the United States has taken on the role of the world's police force, engagements have been progressively bogged down with unrealistic rules of engagement (ROE) and policy-making from people thousands of miles away from the front lines.

In Beirut in the early 1980s, for example, there were several documented incidents of marine snipers being forced to watch atrocities and not being authorized to engage. One sniper watched as "a whole company of unarmed Lebanese soldiers was

⌃ Mil-dot scope reticle: a view not many get to see, and those who do never forget.

massacred in cold blood." Pegler writes: "Another team coming under fire in Beirut in 1983 was denied permission to return fire, as no one could find an officer senior enough to give his authority. Fortunately, other NATO peacekeeping troops serving alongside the Americans did not require such a convoluted system and simply shot back, usually to great effect."

One frustrated marine sniper commented that "We aren't keeping the peace, the Israelis are—they got loaded weapons. The ragheads know they'll get their asses shot off if they pull anything with them."

For their heroism during Operation Continue Hope in Somalia in 1993, two Delta Force snipers were posthumously awarded the Medal of Honor.

Master Sergeant Gary Gordon and Sergeant First Class Randy Shughart volunteered to be dropped into a very hot area to secure the crash site. After pulling the only surviving crew member, pilot Michael Durant, to safety, the two men killed countless Somalis before finally succumbing to the endless tide of militants pouring into the location. Durant survived eleven days in captivity and is alive today because of Gordon and Shughart. Their stories are immortalized in the book and movie *Blackhawk Down*.

Clear on the other side of the character spectrum, there have been some notorious shootings in the United States that some might consider "sniper" attacks, such as Lee Harvey Oswald's shooting assassination of John F. Kennedy in 1963 or Charles Whitman's shooting rampage from the University of Texas (Austin) in 1966. Calling them murderers, psychopaths, or terrorists is far more appropriate. To these men snipers would be to discredit and dishonor the men and women who have performed as genuine snipers in combat.

⋩ **Randy Shughart and Gary Gordon**

Ramadi, Iraq, 2005: A Sniper's Recollections

To conclude this brief history, we asked another former Navy SEAL, Lieutenant Scott Tyler, to share some of his recollections on the experience of being part of a sniper team in Iraq:

Being out at night was always the best time to set up positions and wait for the insurgents to come out in the morning hours to set up their ambushes and improvised explosive devices (IEDs). They seemed to gravitate to the same positions over and over again, usually to spots that were tactically advantageous. (Those insurgents who chose their positions poorly often didn't get a second chance.)

However, while they were able to pick decent positions for setting up on the U.S. military units in the area, the fact that they would use the same positions over and over also provided a pattern that we could exploit.

We were a small number of snipers doing our best to cover a huge area. The cost of failure was high, the rewards were a hot meal and the knowledge that there were a few less terrorists in the world.

When my team first arrived in country, we split up and took different locations throughout the country. My task unit headed west and took responsibility for the Al Anbar province, which included Ramadi and Fallujah. Fallujah was the city where the four Blackwater employees were ambushed, mutilated, burned, and hung from the overhanging beams on a bridge at the city's perimeter.

≳ **Scott Tyler enjoys one last cup of coffee before going into the field.** (*Scott Tyler*)

By the end of 2004, the city of Fallujah had the crap kicked out of it by the Marines in an operation called Vigilant Resolve. It was one of the most dangerous cities in Iraq up to that point, completely lawless and out of control, and was as close as possible to an operational safe haven for local insurgents and Al Qaeda in Iraq.

After Fallujah fell and the terrorists were killed or driven out of Fallujah, the activity seemed to pick up in Ramadi, which was the next city to the west.

When my unit was out there, the Marines and Army were losing several people a week, and sometimes as many as five or more. These were significant losses, considering that an average Marine battalion is only about 500 people, give or take a hundred, and not all of them are out beating the streets on patrols.

▲ **Scott Tyler on walkabout with the locals in Ramadi (*Scott Tyler*)**

The best way to understand the situation on the ground in an area where you are working is to walk it, which is exactly what I and the rest of my sniper detachment did. We would embed with the Marines or the Army, whichever unit had the operational control of the area of operations (AO).

Once we arrived in the area where we were going to operate, one or more of us would head over and talk to the commanding or operations officer, who would give us permission to liaise with their senior enlisted non-commissioned officers, the guys who ran the troops.

From that point we would plan and coordinate patrols, link-ups, break-offs, pick-ups, and contingency actions. Another important step in the process was identifying hot spots, areas where insurgents tended to attack repeatedly. From this information we would start conducting map studies to find areas that we could get visibility on these locations and had a good chance of having a clear shot.

After the preliminary analysis, we would attach to the conventional units and patrol with them during the day to get first-hand looks at potential hide sites, different look angles, and firing positions. I would take several photos from different angles and perspectives and then mark it on my GPS so we could return later and set up shop.

Strict curfews were imposed in Al Anbar during the nighttime hours, and the insurgents knew very well that if they were caught out at night, when most of the citizens were inside, the odds were good that they would be killed or captured, especially if they were caught in the act of setting up an IED or ambush. This meant that in order for them to be effective and increase their chances of survival, they needed to blend in with the local population and try to carry out their nefarious plans while remaining unnoticed by coalition forces.

This situation made nighttime the best for us to move in to get to a location and set up a hide site or firing position. Once in position, it was a long wait till daylight, when our teams would be on hyper-alert, scanning and watching the crowds for anyone doing anything suspicious. Once someone flagged themselves, the burden was on us to positively identify the perpetrator and intended actions and then wreak immediate and final judgment.

2

THE TWENTY-FIRST-CENTURY SNIPER DEFINED

≈ Navy SEAL Chief SOC Brandon Webb

⌃ A small SEAL element trains in mountain operations in Alaska.

In Vietnam, it took the big military machine an average of 2,000 shots to achieve one confirmed North Vietnamese army kill. It took a U.S. military sniper an average of 1.3 shots. Talk about your cost-effective war machine.

After the evacuation of Hanoi in 1975, the United States entered a relative lull in global hostilities, taking part in only a few small conflicts over the course of the next sixteen years. The loss of the experienced war fighter over this span of years brought about a bureaucratic military establishment that eventually forgot about the professionalism, power, and effectiveness that a single well-trained marksman brings to the battlefield. With the terrorist attacks of September 11, 2001, that rapidly changed.

The twenty-first century has seen a shift in training and tactics whose effectiveness is proven immediately in real-time combat. While it would be irresponsible for us to go into detail about modern sniper training methods, it is important for us as former SEAL snipers to provide you with a brief outline of what a SEAL sniper student is up against when his training begins.

Recon operations in Afghanistan. A SEAL sniper must have the conditioning to operate in every environment. ⌄

The U.S. Navy SEAL course is divided into three phases over ninety days, and it tests to the highest standards in the world.

In phase one the candidate learns the latest in digital photography techniques, computer image manipulation and compression, and satellite radio communications. Historically the sniper would sketch a target in detail and record notes with pencil and paper. In the twenty-first century, the sniper leverages technology to his advantage and uses the most advanced digital camera systems, hardware, and software available to record target information and produce a firing solution.

Phase two is the scout portion of training. The name of the game is stealth and concealment. In this phase the sniper learns the art of camouflage, small-unit tactics, patrolling techniques, and, most important, how to get in and out of hostile enemy territory undetected. We often fail candidates who leave behind the smallest trace; a bullet casing left behind will get you sent home.

Toward the end of this phase we introduce advanced marksmanship fundamentals and a system of mental management used by the world's top athletes. When I became involved in rewriting the SEAL Sniper Course in 2003 and 2004, my colleagues and I introduced a wide range of new elements and refinements. Of all the changes we made during these few years, the one that felt most significant to me and that I was personally proudest of was our system for mental management training. (I write about this training and all the other changes we made in the sniper course extensively in my book *The Red Circle: My Life in the Navy SEAL Sniper Corps and How I Trained America's Deadliest Marksmen*.) Mental management gives the students the tools to cope with adversity and a system for rehearsing and practicing their skills perfectly through mental visualization techniques.

To illustrate the value of visualization and mental rehearsal, I often told our sniper students a true story of Captain Jack Sands, a navy fighter pilot who was shot down in Vietnam and imprisoned for years in the Hanoi Hilton, the infamous prisoner of war camp.

Sands was an avid golfer back home, and to get through his extremely demanding situation

⋩ **SEAL sniper student, well concealed and ready to shoot**

he spent hours every day shooting rounds of golf in his head. For years he played his favorite courses perfectly, over and over, in his mind. Eventually liberated and back on U.S. soil, the first thing he did was jump out of the military ambulance and onto the golf course.

After explaining his ragged looks (he was a tall man and quite skinny from malnutrition), he shot eighteen holes of golf at par—a far better score than he had ever done before, prior to his wartime experience. This startled those who witnessed the event, and when questioned about how it was possible, he replied, "Gentlemen, I haven't hit a bad shot in four years!"

Phase three is the sniper portion. We spend hours in the classroom learning the science behind the shot, including ballistics, environmental factors, and human factors, and how to calculate for wind, distance, and target lead. We then put the knowledge to practical application on the shooting range. The students train and test with moving and pop-up targets in high-wind conditions out to 1,000 meters.

As part of the training, we put the shooters in the most stressful and challenging situations imaginable. We look for signs of high intelligence, patience, and mental maturity. Then we place the shooter in adverse and unfair situations, often without their realizing it, in order to test their mettle.

An example of this would be the "edge" shot. Individual trainees are lined up on the shooting range and are told they have four minutes to run 600 meters, set up on the firing line, and wait for their targets to appear, which could be at any time between four minutes and one second to an hour later. But we would always send a target up right at three minutes—usually right as the shooter was still setting up on his lane and identifying his field of fire!

⌃ **Some interop sustainability training. If you want to stay sharp, time on the trigger and behind the spotting scope is mandatory.**

Often a shooter will take his eyes away for a split second to wipe sweat from his brow, then drop back down on his scope just in time to see his target disappear and his opportunity to score with it. The peer pressure is intense and shooters often break down in frustration at a missed shot. They eventually learn to control their feelings or they don't move on. As instructors we record everything and keep detailed student records.

∧ Operating from an MH-6 little bird will test the skills of even the best sniper.

Just getting a billet in the SEAL Sniper Course is extremely competitive, and even then, a large percentage of candidates don't make it through. No one wants to go back to his SEAL team deemed a loser for having failed out of the course. However, this course is one of the few you can fail as a SEAL and not be looked down upon by your teammates. This is because the SEAL Sniper Course is renowned for being one of the toughest and most challenging courses in the world.

More than three months of seven-day, hundred-hour workweeks go into the training. It takes exceptional

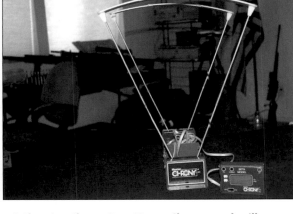

∧ A Shooting Chrony Beta Master Chronograph will accurately measure the muzzle velocity for a particular round and rifle combination, data which will produce an unbelievably accurate first round . . .

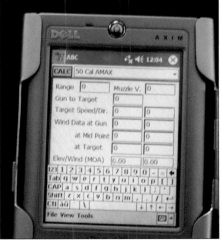

∧ . . . when input into a device like this PDA— though technology is starting to push this data into devices that also double as tactical watches.

⊼ **SEALs practice a GOPLAT (Gas & Oil Platform) takedown. SEAL snipers will often support in the helicopter.**

perseverance in order to graduate with the title of SEAL sniper. It remains one of the most stressful events of my life—even when compared to my combat tours.

The training that takes place in modern military-grade sniper communities is far more advanced than anything seen before in the history of military training.

One major shift in today's training methodology is that modern-day Special Operations snipers are trained and deployed as independent shooters, rather than only in the traditional shooter/spotter pairs.

Past training courses would produce some graduates who were great shots but not so great behind the spotting scope, and vice versa. Tough as it is to admit, we were graduating students who were not well-rounded snipers. Because students were trained in pairs and shared scores, often they would make up for

⌃ **American Sniper Jack Murphy conducting operations with Afghan National Army**

each other's deficiencies, and this is not a good way to conduct business or create a world-class sniper.

The truth is, 90 percent of our actual field sniper missions in the SEAL teams are conducted with single shooters in separate positions.

When I became the course manager, I pushed our officer in charge to accept that if most of our snipers were being used independently, then we needed to start training the way we fight and focus on graduating a well-rounded sniper capable of operating on his own, without a spotter. The first time a sniper operates independently should happen in training—not on a real-world mission!

From that point on we gradually shifted the focus of our training and testing to ensure that our SEAL students had a complete grasp of all the subject matter.

My prediction is that modern courses will continue to shift toward training their students more in the use of technology, as well as graduating snipers capable of deploying as single shooters who don't require or rely on a spotter. The modern graduate can deploy in any combat theater without the aid of a spotter because he has the training and a complete suite of technology at his fingertips. The days of dope books and hand sketches are falling to the wayside and being replaced by digital imagery, handheld computers with complex ballistic software programs, chronographs that measure each weapon's specific muzzle velocity (two identical rifles shooting the same round will produce different results), nanotechnology applied to camouflage, and extremely accurate rapid-fire smart weapons.

The twenty-first-century sniper is a mature, intelligent shooter who leverages technology to his deadly advantage.

He has spent thousands of hours honing his skills. He is a master of concealment in all environments, from the mountains of Afghanistan to the crowded streets of Iraq. He is thoroughly trained in science, but he alone is left to create the individual art of the kill.

To the sniper, the battlefield is like a painter's blank canvas. It is up to him to use his tools, training, and creativity to determine how that final shot will play out, creating the devastating psychological impact that is the ultimate result of his actions.

3

THE CLASSIC: BOLT-ACTION RIFLES

The world today is truly a dangerous place. Urban crime, the global "war on terror" (which is not really a war against *terror* at all, but a war against groups holding specific radical philosophies), homegrown terrorists plotting to kill us, the wars in Iraq and Afghanistan, desperate men holding innocent people hostage . . . all these are evidence of just how dangerous our world has become.

In every one of these violent situations there is a leader, someone who is directing forces or holding hostages and demanding terms.

This is a critical principle that applies to any endeavor, be it a business, a military unit, a pirate crew holding a ship captain hostage, or a crazed man holding his family hostage: Once you eliminate the leader, the rest of the organization will quickly falter.

All the violent situations described above call for the highly refined and unique skills of the twenty-first-century sniper.

The capabilities of the modern-day sniper are vast. Every job requires the same basic components: someone who is motivated to do the work, possesses the necessary skills and competence to perform the task, and is provided with the proper equipment to complete the job. If any of these three components is lacking, the job will be performed inadequately or not be accomplished at all. Without the proper tools, in some cases you can get the job done, but only very inefficiently. You could use a screwdriver to nail a railroad spike, you could use a steak knife to cut down an oak tree, or you could use a Swiss Army knife to attack a charging grizzly, but the outcome may not be what you were hoping for.

The twenty-first-century sniper uses a wide range of specialized tools and technologies to accomplish his

⌃ **Glen poses on the set of a *National Geographic* sniper special in southern California.**

varied missions, and some rifles and calibers have broader application than others.

If the sniper's mission calls for eliminating a gang member holding a woman hostage in a bank, the use of a .50-cal weapon would get the job done, but it would be overkill: a .50-cal round will likely continue ripping through multiple unknowns after taking out its intended target, including the back wall of the bank. Likewise, if the mission is taking out an enemy combatant at 1,500 meters, a .308 would not be the ideal tool of choice.

Later in this book we'll explore comprehensive issues that relate to all sniper rifles, but first we'll take an in-depth look at the two principal types of weapons snipers use, along with their advantages and disadvantages. In this chapter we look at the bolt-action rifle, and in chapter 4, the semiautomatic rifle.

A Brief History of the Bolt-Action Rifle

Bolt-action rifles have been in use since the middle of the nineteenth century, and while they are being slowly phased out by the increased accuracy, practicality, and reliability of the semiautomatics, they remain in heavy use today.

The bolt action came into existence in the mid-1800s along with the development of smokeless powder. Prior to the advent of smokeless powder, rifle ammunition was powered by black powder. As most shooters know, black powder produces a large volume of smoke when ignited—not a good thing if you are a sniper, trying to fire upon the enemy without being detected.

Johann Nikolaus von Dreyse designed the first crude bolt gun in 1848, and in 1866 the French introduced their version of an early bolt rifle, the Chassepot "Needle Gun," officially known as the Fusil Modèle 1866. These new rifles gave the sniper the ability to fire a round by releasing the firing pin, which allowed the trigger to be lighter, thereby giving the sniper better trigger control.

In 1871, the German brothers Paul and Wilhelm Mauser introduced the first true bolt-action rifle, which was the first to use a true center-fire cartridge. At the time, the U.S. Army was using the .45–70 Government (.45-cal bullet with 70 grains of black powder) that generated about 1,300 fps (feet per second) at the muzzle. The French-designed 8 x 50mm

Friends at the range. L to R: Billy Tosheff, Brandon, Kevin Kouzmanoff, and a SEAL sniper.

⌃ Chassepot bolt and chamber with old round

⌃ Modern bolt-action rifle: the CheyTac M200 Intervention .408 (CheyTac)

Lebel using a 198-grain bullet generated 2,380 fps at the muzzle, really the first smokeless round to be widely used by any one country.

Across the English Channel, the British introduced the Lee-Enfield 303 in 1888. This bolt-action rifle was capable of using metallic cartridges and could be loaded with more than one round in the chamber—one of the first repeaters in that it had a ten-round magazine. That same year the Mauser brothers introduced their next-generation bolt rifle, the 8mm Mauser, to the German army. The action on this rifle is still used today.

A few years later the Russians developed the Moisin-Nagant Model 1891, which fired a new rimmed cartridge, the 7.62 x 54mm. Like the Mauser for the Germans and the Enfield for the British, the Moisin-Nagant saw service through World War II as Russia's primary sniper rifle.

By the late 1880s, the United States recognized that its single-shot .45–70 Trapdoor Springfield was not a suitable rifle, given developments in Europe.

In 1890 and 1891, the U.S. Army looked at many rifles and eventually decided to go with a rifle designed by Norwegian Army Captain Ole Krag and Erik Jørgensen—the .30–40 Krag, which was designated as

Mauser 98k with Zeiss optics ⌄

the Model 1892. This rifle shot a .30-caliber 220-grain bullet with 40 grains of smokeless powder. This load combination generated about 2,000 fps at the muzzle, which was inferior to the German, French, and British cartridges but better than the .45–70.

The effective range of the Model 1892 was about 360 to 450 meters, but the best group size was about eight to ten centimeters at 90 meters—a horrible 3.5–4 MOA, or minute of angle. (Minute of angle, also called minute of arc, is the common unit for measuring firearms accuracy.)

During World War I the German Mausers were the best sniper-specific rifles in existence. The Germans were also the first to put optics on their sniper rifles.

In the United States, meanwhile, the .30–06 Springfield, which was adopted in 1903 along with the Model 1903 Springfield, had become the standard U.S. military cartridge, and was designated as the .30-03. However, severe erosion of the 1903 Springfield barrel's throat with the powder being used at the time forced the round load to be reduced, which resulted in a muzzle velocity of only 2,000 fps—the same as the .30-40 Krag that the original .30–06 was supposed to replace.

U.S. Ordnance modified the cartridge case slightly with a 150-grain bullet, which then generated a muzzle velocity of 2,700 fps. The cartridge was then designated the .30-06, which saw service throughout both World Wars and is still in use today.

Dimensions of the modern-day round ≈

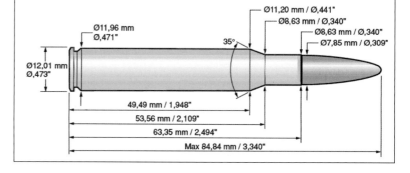

Development of the .308 began at the end of World War I, but it was not until 1957 that the government finally used the 7.62 NATO round in more than one weapon, getting it into the M14 and M60 machine gun.

As developments in metallurgy, powders, bullets, and the science of projectile ballistics continued to progress, other cartridges were developed to provide the sniper with better tools to accomplish his mission. The .308 remains a very popular cartridge for the twenty-first-century sniper, in both bolt-action and semiautomatic versions. It is probably the most accurate .30-caliber commercial cartridge ever produced.

The Components: How the Bolt-Action Rifle Goes "Bang"

Rifles—all guns, for that matter—can be classified by the type of action that places the cartridge into the firing chamber. Modern rifles have three different actions: a) a lever action, b) a semiautomatic action, and c) a bolt action.

In the lever-action rifle, a lever pulls and pushes the cartridge into and out of the chamber. Lever-action rifles are commonly depicted in Westerns and are often referred to as "cowboy rifles."

Semiautomatic rifles, such as the M16 and M14, use springs along with the gas generated by the burning powder to move the loading/unloading mechanism.

In both lever-action and semiautomatic rifles, the loading/unloading mechanism may be referred as a "bolt" in the owner's manual, and it does in fact perform the same function as the bolt in a bolt-action rifle.

The Marlin lever-action rifle (*Benelli USA Corp.*) ⌄

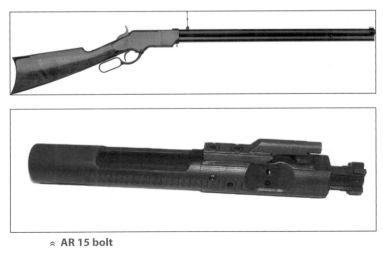

⌃ **AR 15 bolt**

In a bolt-action rifle, the bolt is moved in and out of the receiver by the shooter's hand.

To load the cartridge into the firing chamber, the bolt is pushed and twisted forward. To unload and eject the spent cartridge, the bolt is twisted and pulled back. This twisting motion is similar to the action used to tighten and untighten a bolt, hence the name *bolt-action*.

All rifles are composed of the same four basic components: a) the action; b) the trigger/safety mechanism; c) the barrel; and d) the stock.

The *action* is composed of two parts: the receiver and the bolt. The receiver is where the cartridge enters the action. The bolt pushes and pulls the cartridge into and out of the firing chamber.

The bolt houses two components: the firing pin, which strikes the primer of the cartridge when the trigger is pulled, and the extractor, which pulls the spent round out of the chamber and ejects it in preparation for the next cartridge.

The action plays an integral part in the rifle's firing. In sniper-grade rifles, the action parts are machined to very close tolerances and should be trued, as even minor misalignments can cause problems with the accuracy of the rifle over long distances.

The *trigger mechanism* is made up of several parts that work together to deliver a strike or hammer to the firing pin, located (usually) in the bolt. The hammer drives the firing pin forward to the primer of the cartridge.

When the firing pin strikes the primer, the explosive component in the primer is ignited, which in turn ignites the powder in the cartridge case. As the powder burns, it very quickly creates an enormous amount of gas that expands rapidly, forcing the bullet in the case down the barrel and out to its intended target.

There are two types of triggers: two-stage and single-stage. We address the differences between them in

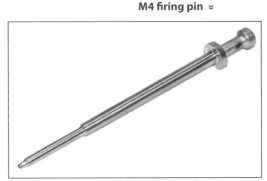

M4 firing pin

Barrel rifling: a bullet's-eye view

chapter 6. In sniper-grade rifles the trigger must be very smooth, dependable, safe, and easy to manipulate. A smooth trigger mechanism is a key component in building an accurate sniper rifle.

The *barrel* is probably the most important factor influencing a rifle's inherent accuracy. Barrels and their characteristics will be discussed in greater depth in chapter 5.

The barrel begins with the chamber, into which the bolt pushes the cartridge in preparation for firing. The chamber then narrows to the designated caliber diameter for the length of the barrel.

All rifle barrels have spiral grooves, known as *rifling*, cut into them that engage the bullet as it travels from the cartridge to the end of the barrel. The spiral curve of the rifling causes the bullet to spin, which imparts a gyroscopic stability that greatly increases the projectile's accuracy. The science and art of building an

McMillan stock (*Curtis Prejean, courtesy of McMillan*) »

Modern-day bolt action; notice the adjustable cheek weld and rail system for mounting additional lasers and optics. »

accurate barrel is complex, as many factors influence the barrel's accuracy.

The *stock* is the part of the rifle that allows the sniper to hold on to the rifle. This can be made of wood, fiberglass, aluminum, or a combination of these materials. On a modern sniper rifle the stock is usually made of fiberglass and synthetic material or aluminum.

In addition to these four components, there are other accessories that must be added to the rifle to make it functional. The first mandatory accessory is some kind of sighting device. Sight options can vary from the basic iron sights to optical telescopic sights. Modern sniper rifles are always equipped with optical telescopic sights, so discussing iron sights in connection with

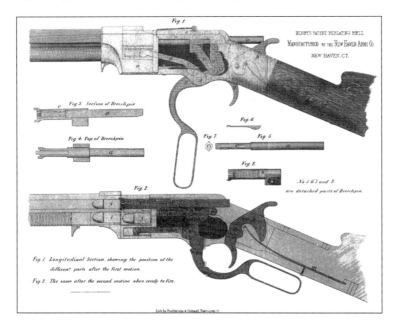

« **Details of 1860 Henry rifle (*Benelli USA Corp.*)**

bolt-action sniper rifles is pointless—they don't have them. Rifle scopes and optics are discussed in chapter 9.

Optional sniper rifle accessories include a muzzle brake, a bipod, and perhaps a suppressor. Suppressors are discussed in chapter 8.

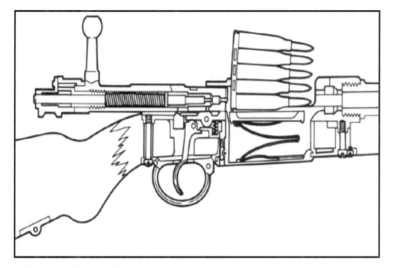

⌃ Cutaway view of the Accuracy International .244 (*Curtis Prejean, courtesy of Accuracy International*)

McMillan action explained (*Curtis Prejean, courtesy of McMillan*) »

Actions

MCMILLAN ACTIONS.
PRECISION ENGINEERING IS THE HEART AND SOUL OF THE FINEST RIFLE YOU WILL EVER OWN.

A McMillan action represents the finest ideas from the laboratory of the long-range competition circuits combined with the proving ground of actual combat and law enforcement experience. Precision tolerances. Superior steel. Thoughtful ergonomics. And styling like a sleek sports car. A McMillan action is the foundation for a true professional's rifle.

FLUTED BOLT AIDS RELIABILITY.
Flutes sweep mud and grit away, helping to keep the action operating smoothly in harsh environments.

RECESSED BOLT FACE FOR STRENGTH.
The bolt face, the barrel shank and the action all surround the cartridge head to add strength in this critical area during ignition.

McMillan standard action

ACCEPTS REMINGTON-STYLE TRIGGER.

MODEL G30

0709001

EVEN THE UNDERSIDE OF THE McMILLAN ACTION IS A WORK OF ART.
In a production action, this area is finished poorly with burrs and tool marks to save costs. In contrast, every component of a McMillan action shows precision engineering and attention to detail throughout, whether the area is hidden or exposed.

TIGHT BENCHREST TOLERANCES.
Decades of experience in the benchrest circuit have shown that absolute concentricity of the bolt and the action body around the axis defined by the bore line of the barrel is paramount to precision shooting. Each McMillan action is CNC machined from a centerless ground billet blank, not a casting. The face of the action is absolutely perpendicular to the axis of the bolt. Tolerances on general metalwork are held to within 0.005". Critical parts are held to 0.0005" tolerances.

« Cutaway of an Accuracy International bolt action (*Curtis Prejean, courtesy of Accuracy International*)

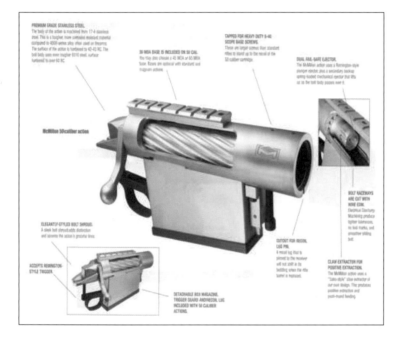

« McMillan .50-caliber action (*Curtis Prejean, courtesy of McMillan*)

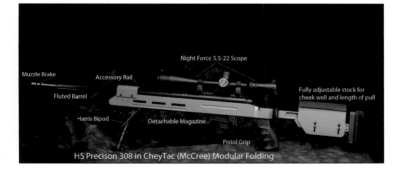

« HS Precision .308

4

SEMIAUTOMATICS: THE WAY OF THE FUTURE

Semiautomatic: a *firearm*, able to fire repeatedly by automatically ejecting the spent cartridge case and loading the next cartridge from the magazine, but requiring release and another pressure on the trigger for each successive shot.

A (Very) Brief History of the Semiautomatic Rifle

There were a few early attempts to create a rifle that didn't require manual reloading after each shot, but it was not until 1885 that a German gunsmith named Ferdinand von Mannlicher invented the Model 85, the first known successful semiautomatic rifle. American John Browning followed up a few years later with the first successful semiautomatic shotgun, the Browning Auto-5.

The Auto-5 relied on long recoil operation. This design coupled the recoil action with a spring to rechamber the next round, and it was in military service for over fifty years. (Production of the Browning Auto-5 continued until the end of the twentieth century.) Winchester was the first to introduce the semiautomatic rimfire and centerfire rifles into the civilian market in 1903 and 1905.

Surprisingly, the first country to incorporate a semiautomatic rifle in its military service was France, with the Fusil Automatique Modèle 1917. This was a recoil-operated action rifle used in World War I.

Toward the latter part of World War I, the Soviet Union, Germany, and Britain had all started incorporating the semiautomatic into their inventories. Britain's goal was to replace the very reliable bolt-action Lee-Enfield rifle. I have personally run across a few of the Enfields still in use in Afghanistan, a tribute to their solid construction and reliability!

Fusil Automatique patent drawing ⩗

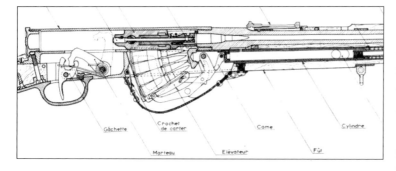

The first semiautomatic that became standard issue for the U.S. military was the M1 Garand, a gas-operated rifle developed in Canada by John Garand for Springfield Armory. The first production model was released in 1937.

The semiautomatic proved itself under fire in World War II and gave the U.S. soldier quite an advantage over German opponents, who were still mostly using bolt-action rifles. It is a bit ironic that the country that invented the semiautomatic didn't incorporate it earlier into their military in larger numbers.

The 1917: a weapon ahead of its time ⩗

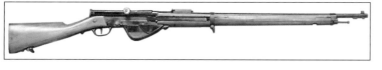

In 1945, as World War II was winding down, design began on a new rifle in the Soviet Union by arms designer Mikhail Kalashnikov. The Soviets are known for their rugged and virtually bulletproof engineering standards, whether it's airplanes, tractors, cars, or guns, and the Avtomat Kalashnikova, or AK-47, is arguably the most popular and widely used semiautomatic rifle today.

⩘ M1 Garand: a battle-proven rifle

Captured selection of semiautos in the early days following 9/11 in Afghanistan ⩘

The AK-47 will stand up to the harshest environments. Whether it's the frigid mountains along the Afghanistan/Pakistan border or the hot, dusty streets of southern Iraq, this weapon will fire consistently and without malfunction. I can personally attest to having used this weapon in training over an eight-hour period, swimming it back and forth through heavy ocean surf and still having the weapon fire flawlessly. Its rugged and simple engineering is why this weapon is one of the most popular in the world.

Paradigm Shift

Bolt-action rifles are still employed in modern warfare, but I am here to tell you that the bolt-action rifle is quickly becoming a thing of the past, much like a manual transmission in an automobile. While some people will continue to prefer the manual over the automatic, it is clear that the old stick-shift is fading quickly, while automatic is here to stay. The same can be said of today's semiautomatic rifles: They are just as reliable and

⩘ The AK, weapon of choice, is shown here with the transportation of choice in more countries than any other, as friendlies get ready to roll in Afghanistan.

≈ **Mural from Iraq: lovely lady sporting the everpresent AK-47, a widely used and reliable weapon**

Modern-day semiautomatic rifle ≈

accurate (sub-MOA, that is, accuracy of less than one minute of angle) as a good bolt-action—and a hell of a lot more practical.

That said, there has to be some discussion of quality. Not all semiautomatics are created equal, and there is plenty of garbage out there.

For a while, in the SEAL teams, we were using the Knight's Armament 7.62 SR25. To be honest, I am not a big fan, as I've personally experienced critical malfunctions with these weapons and seen more than a few of them have total mechanical failure in training. When you are in an urban hide in Basra, Iraq, surrounded by bad guys who want nothing more than to string you up by the neck, having a malfunction is unacceptable.

It is quite unfortunate that there are very few American weapons manufacturers who are currently making a full-scale production model semiautomatic that can compete in the global market.

As with the collapse of the American auto industry, the harsh reality is that few U.S. weapons manufacturers are making high-quality products and pushing the technology envelope. The Europeans have been kicking our asses with their advanced engineering and forward thinking.

I will say that Smith & Wesson Holding Company is beginning to catch on, and in my conversations with management, they say they are positioning themselves for change in the new century. This is refreshing to see from a U.S. company.

⩙ "Only dropped once." Note the blood on the foregrip and the two bullet holes through the weapon, one on each grip. Taken off a fallen Iraqi soldier outside Tikrit, Iraq.

A highly modified semiauto with U.S. Optics SN series variable power scope. A 7.62 that can reach out and engage multiple targets quickly the way only a semiauto can. Note the match-grade barrel shown next to a box of LeMas Ltd's BMT 7.62 match ammo. ⩘

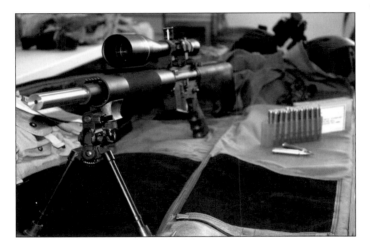

Snipers in today's warfare value the semiautomatic for its reliability and accuracy, but also as a weapon that can engage the enemy in a sustained firefight and also serve as a battle weapon in close quarters, such as when clearing multiple rooms in a building.

There is no substitute for the versatility and adaptability of the modern-day semiautomatic. In Iraq our SEAL sniper teams were devastatingly effective with the semiautomatic. We owned the urban environment and literally shut down miles of city with the rapid accurate fire our semiautomatics were able to deliver. We have some of the highest kills in the Special Operations community because of our high training standards coupled with the modern semiautomatic.

Semiautomatic Operating Systems

Modern semiautomatics fall into three categories of operating system: blowback-operated, recoil-operated, and gas-operated.

Brandon checks out the feel of the suppressed HK 417.

In most blowback-operated designs, the bolt is unlocked at the moment of firing. There is a delay mechanism used to prevent excessive force from operating the bolt; your kid's Ruger .10/22 is an excellent example of this action. For the grown-ups, the H&K PSG-1 uses a roller-delayed blowback design to completely close the bolt face prior to firing, increasing the accuracy to sub-minute of angle and increasing the range to 800 meters.

Heckler & Koch PSG-1 counter-sniper rifle *(Heckler & Koch)*

The roller delay mechanism has been around for decades and was first deployed by Mauser in 1945. The mechanism forces the bolt carrier rearward at high velocity and leaves the bolt face in place, letting the mass of the bolt provide inertia that creates a time delay required to ensure that the bullet has left the barrel and the gas pressure in the barrel has dropped to a safe level.

It's still unpleasant to stand next to. If the hot casing doesn't hit you, the gas from the chamber will. Many sniper-quality semiautomatics fall into this category.

In recoil-operated weapons, the bolt is not completely engaged and is held closed by a spring. When the weapon is fired, the recoil overcomes the spring tension and the bolt can cycle. The spring tension decreases with use over time, and as the force on the cartridge decreases, the force on the bolt face becomes inconsistent and the effect on the bullet trajectory is drastic.

Recoil operation is usually limited to smaller caliber weapons, due to the amount of force transferred to the weapon frame and the operator by the moving bolt.

Gas operation of self-loading firearms has been around for more than a century. The early designs were often prone to failure. The first design that was considered usable was John Browning's in 1891, with the first working prototype capable of firing sixteen .44-caliber rounds in under one second.

There are two types of gas-operated mechanisms. The first uses a very stiff spring to keep the bolt face closed; a small amount of gas is tapped from the barrel to push the bolt back against the spring. The AK-47 uses this method, which has the advantage of being very inexpensive and simple to manufacture.

The second type of gas-operated semiautomatic operating system is the direct impingement method, in which the tapped gas is vented into a tube on the bolt directly and blows the bolt backward after firing, cycling the weapon. The M16 uses this method of

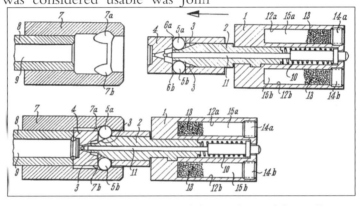

≈ HK is famous for the reliability and the simplicity of their roller-delayed operation, shown here in detail.

A traditional shooter/spotter pair using an M21 on the left and an M24 on the right. Oftentimes a spotter will use a rifle scope to spot for a shooter at closer distances. (*Department of Defense Visual Information Center*) ≈

operation with the addition of a rotating bolt to increase accuracy.

There are a large number of variations, such as the tilting breechblock used by Fabrique Nationale de Herstal (FN) in their FAL weapons to create a consistent bolt-face closure and increase accuracy and range. The FN's Special Operations Combat Assault Rifle (SCAR) uses a gas-operated rotating bolt to achieve a respectable cycle rate of 625 rounds per minute, with sub-MOA accuracy and 800-meter range for the 7.62 x 51 NATO cartridges.

Two Standout Semiautomatics

Two great examples of modern sub-MOA semiautomatics are Heckler & Koch's MSG90A1 and the FN's Special Operations Combat Assault Rifle (SCAR light and SCAR heavy) system, which has been selected by U.S. SOCOM for use by all U.S. Special Operations forces.

Heckler & Koch is a German weapons manufacturing company that is perhaps most famous for its MP5 submachine gun, commonly referred to as the "room broom" by Special Operations personnel. H&K was the first arms manufacturer to use synthetic polymers in their weapons, which reduced weapon weight significantly.

The MSG was Heckler & Koch's answer to a militarized version of the PSG-1 and is a very reliable, finely engineered

≈ **A SCAR heavy with ACOG**

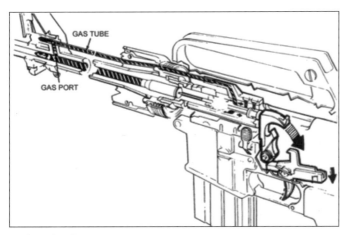

≈ **Direct impingement gas pathways can lead to a lot of fouling and carbon buildup in the chamber.**

subminute gun (that is, a very accurate weapon capable of shooting sub-MOA groupings). The barrel is weighted precisely at the muzzle to increase the harmonic stabilization of a distortion factor called barrel whip, which increases the accuracy. We'll look at barrel factors (including barrel whip) in more depth in chapter 5.

I can honestly say that in over a decade in the SEAL teams, I have never had a Heckler & Koch weapon malfunction on me.

The FN SCAR system offers the user the option to switch caliber with ease. That is invaluable because as it simplifies things greatly and gives the sniper choices both urban and rural. The company produces a SCAR light (5.56) and a SCAR heavy (7.62) version.

▲ **HK MSG90A1 (***Heckler & Koch***)**

After achieving great success in 2007 in a limited competition among the M4 Carbine, the FN SCAR, and the previously discontinued H7K XM8, the FN SCAR-L replaced the M4 Carbine and the FN SCAR-H replaced the M14 and MK11 sniper rifle.

The main advantage of the FN SCAR system is a huge increase in the capacity to deliver rounds on target. When shots are snapping over and around you in the heat of battle, this ability to take down targets in very rapid succession is critical. In addition, it provides a versatile weapon that can be used in close-quarters battle and helicopter-sniper operations. Couple this with modern reliability and you can say good-bye to the bolt-action rifle.

FN SCAR with Nightforce optics: a great combination ▼

No doubt there are some hard-core holdouts who will insist this is not the case and that nothing takes the place of the bolt-action. But this simply isn't true.

⌃ **The FN F2000 series is a well-balanced carbine that can be tailored as the primary weapon for a variety of missions. Not exactly sniper rifles, but definitely an innovative shift in weaponry and a good urban sniper sidekick.**

I would ask these holdouts, "Have you had any recent experience on the asymmetric modern battlefield?"

I didn't think so.

When I was in northern Afghanistan conducting reconnaissance and patrol missions, I was torn between the reliability of my .300 Win Mag and the comprehensiveness and rapid-fire capability of my SR–25 semiauto. The problem was, I didn't trust my semiauto to hold up to the harsh conditions because I'd had malfunction problems with the Knight's Armament weapon in the past.

Ultimately, I chose to patrol with my M4 and carry my .300 Win Mag on my back. This is the reason bolt-action rifles have long been the sniper's choice: You simply trust the reliability of the bolt-action system.

With the reliability and accuracy of today's semiautos, I would choose a semiauto in this situation, hands down. I would have been lighter on my feet and had a much more efficient weapon system.

The modern semiautomatic sniper has proven its accuracy and reliability on today's battlefields over and over again. It has performed reliably and taken down thousands of targets on the complex urban environments of Iraq and in the freezing mountain terrain of northeastern Afghanistan. It will only be a matter of a few years before the bolt-action rifle is phased out altogether.

The semiautomatic has earned its rightful place in the inventory . . . until the next innovative shift in weaponry occurs.

⩘ **Brandon training with Kurdish commander in northern Iraq**

A CH-53 comes in to extract Brandon's SEAL platoon from the mountains in Afghanistan. ⩙

5

RIFLE SCIENCE: WHAT EVERY SHOOTER SHOULD KNOW

Any discussion of what makes a good sniper rifle, of what factors determine whether it will serve reliably in the field or run the risk of crapping out and letting you down, rightfully starts with the largest and in many ways most critical component: the barrel.

A nonshooter will look at a rifle barrel and see a round metal bar with a hole drilled down the center. He will probably realize that this is the part of the rifle where the bullet leaves the rifle. However, the science involved in building an accurate sniper rifle barrel is vastly more subtle and complex than simply drilling a hole through the length of a metal bar.

The barrel's length, the type of steel or material used, the rifling, the twist rate, and the weight all play very important roles in the accuracy of the barrel. In fact, the barrel is probably the most important factor in determining a rifle's inherent accuracy. A basic knowledge of what makes a good barrel will help you understand what it takes to build a great rifle, and understanding a barrel's characteristics will not only help you appreciate what your rifle can do and why, but it will also help you diagnose any accuracy issues that might arise.

The Secret of Steel

"The secret of steel has always carried with it a mystery. You must learn its riddle, Conan."—Conan's father, in *Conan the Barbarian*

Today most rifle barrels are constructed of some kind of steel alloy. Steel alloys contain elements that are added to modify the steel's behavior during the heat-treatment portion of manufacturing. Examples of these added metals include nickel, which increases hardness and tensile strength; vanadium, which improves the ability to resist repeated stress; tungsten, which adds air-hardening qualities; molybdenum, which improves resistance to high temperature; and chromium, which imparts added stiffness and hardness to the alloy.

The most common steel material in rifle barrel building is known as chromoly. This steel alloy is very high in strength, if manufactured correctly; its tensile strength can vary from 98,000 pounds per square inch (psi) to 180,000 psi. Chromoly also con-

Range time on a .50. Marksmanship is a perishable skill requiring continual training and refinement. »

tains chromium, manganese, molybdenum, phosphorus, sulfur, and silicon, but no nickel.

The other type of steel used in rifle barrels is stainless steel, which contains at least 10.5 percent chromium and is more expensive than chromoly.

Patented in 1916 and manufactured by the American Stainless Steel Corporation, stainless was originally used for knives, but the name was not trademarked, so today the "stainless" designation is used for any steel that resists rust, corrosion, and acids. There are over fifty different stainless steel compositions, all with their own unique properties.

The tensile strength of stainless is lower than properly processed 4140 chromoly steel, and it's worth noting that forging an alloy with the tensile strength required for rifle barrels also results in a steel that will rust under some conditions. Stainless is good at reducing barrel wear and fouling.

In the manufacturing process, the steel goes through a series of heating and cooling stages to condition the steel and optimize its performance. A good steel barrel should be stress-relieved during the manufacturing process, a process of heating and then cooling the barrel in a controlled way. This allows the steel molecules to relax and align themselves, which reduces any imperfections in the trueness of the barrel.

Once the steel blank is properly processed, the barrel must be drilled, or reamed. Usually the reaming process takes several passes to bring the barrel to the correct diameter. The outside of the barrel has to be machined to the correct size and should be precisely concentric to the bore of the barrel.

A "gang broach" for rifling barrels. This method of cutting the rifling into barrels is effective, but making the broach and keeping it sharp are difficult machining operations. ≈

Beautifully rifled shotgun barrel; most shotguns are smoothbore. ≈

≈ **Rock River Arms Tactical CAR A4 with a sixteen-inch chromoly barrel (*Rock River Arms*)**

The rifling process (cutting of the spiral grooves on the interior) can be accomplished in several different ways: hammer forging, button-cut rifling, and single-cut rifling.

Hammer forging involves heating the reamed barrel and then inserting a mandrel, a very hard rod fashioned with spiral grooves, like a drill bit. The barrel is hammered into shape around the mandrel, which leaves an imprint of grooves when it is removed. Hammer forging is usually considered the least accurate method of rifling, but it is also the cheapest, and several manufacturers employ this method in all of their rifles with good accuracy.

In the process of button-cut rifling, a barrel blank is center bored and a hard metal button with the rifling ridges is pulled through the barrel, producing the rifle lands and grooves. Button-cut rifling is more expensive than hammer forging and in most rifle-building circles is also thought to be more accurate.

Single-cut rifling, also known as hook rifling, involves the same center-boring process, but the rifling is cut slowly with a single groove tool. Obviously, this process is very time-consuming, since it takes multiple passes, one for each groove. This process causes minimal stress to the barrel and, if done properly, is thought by many to result in a very accurate barrel. It is also the most expensive rifling process.

≈ **Jack taking a break. Note the excellent camouflage.**

Size Matters

The barrel's length is also an important consideration in building a sniper rifle.

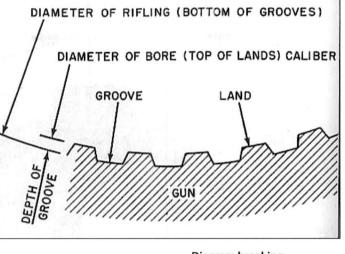

Some factors regarding barrel length are obvious. For example, a long barrel is more cumbersome to handle in confined quarters, is heavier to carry on long missions, and may make shooting in anything but a supported or prone position impractical. On the upside, longer barrels are generally more accurate than shorter barrels, especially at longer ranges.

Diagram breaking down rifling grooves and lands; the caliber of the barrel is measured from the internal diameter, starting with the top of the lands.

The improved accuracy with a longer barrel is secondary to the added weight—less recoil, less movement when the rifle is fired, and higher bullet velocities that result in a flatter trajectory.

The higher velocities produced by longer barrels are generally true—*generally*, not *always*.

Interestingly, despite the fact that shorter barrels produce less bullet velocity, the drop in speed and accuracy is small enough that it is of little concern to most hunters. Hunters who seldom shoot at distances of more than 100 to 150 meters will not notice any discernible drop in accuracy or velocity in a 20- to 22-inch barrel as compared to a 24- to 26-inch barrel. The maximum velocity for any barrel length will vary with the powder, bullet, cartridge, and other factors. After leaving the muzzle of the barrel, the bullet immediately begins to lose speed; hence its greatest speed occurs directly at the muzzle.

When the powder in the cartridge ignites, the resulting hot gasses expand eight hundred to thirteen hundred times. Since the gas is contained in the brass case, there is only one way for

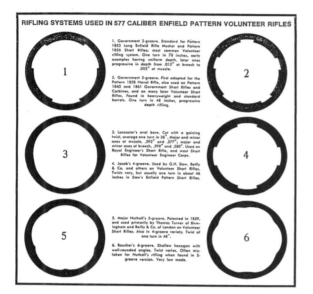

RIFLING SYSTEMS USED IN 577 CALIBER ENFIELD PATTERN VOLUNTEER RIFLES

1. Government 3-groove. Standard for Pattern 1853 Long Enfield Rifle Musket and Pattern 1856 Short Rifles; most common Volunteer rifling system. One turn in 78 inches, early examples having uniform depth, later ones progressive in depth from .013" at breech to .005" at muzzle.

2. Government 5-groove. First adopted for the Pattern 1858 Naval Rifle, also used on Pattern 1860 and 1861 Government Short Rifles and Carbines, and on many later Volunteer Short Rifles. Found in heavyweight and standard barrels. One turn in 48 inches, progressive depth rifling.

3. Lancaster's oval bore. Cut with a gaining twist, average one turn in 36". Major and minor axes at muzzle, .593" and .577"; major and minor axes at breech, .593" and .580". Used on Royal Engineer's Short Rifle, and most Short Rifles for Volunteer Engineer Corps.

4. Jacob's 4-groove. Used by G.H. Daw, Reilly & Co. and others on Volunteer Short Rifles. Twists vary, but usually one turn in about 46 inches in Daw's Enfield Pattern Short Rifles.

5. Major Nuthall's 5-groove. Patented in 1859, and used primarily by Thomas Turner of Birmingham and Reilly & Co. of London on Volunteer Short Rifles. Also in 4-groove variety. Twist of one turn in 48".

6. Boucher's 6-groove. Shallow hexagon with well-rounded angles. Twist varies. Often mistaken for Nuthall's rifling when found in 5-groove version. Very few made.

⌃ **Old chart showing different rifling patterns for the Enfield rifle**

Brandon behind a LaRue Tactical ⌄

the gas to escape—by pushing the bullet down the barrel. The burning of the cartridge powder should occur in the cartridge case, not in the barrel—and the barrel should have nothing in it except the bullet and the push of expanding gas.

The bore of the barrel should be as smooth as possible. The reaming and rifle-cutting process will leave small irregularities that need to be removed by smoothing the barrel with a lapping process and/or going through the break-in process. The break-in process varies with manufacturers but usually consists of firing a certain number of bullets while cleaning the barrel between rounds. This process helps smooth out the small imperfections left by the reaming and rifling process.

Regardless of the strength of the steel used, firing high-velocity bullets will cause all barrels to wear out eventually. As the barrel wears, accuracy deteriorates. As the barrel ages, the shooter will notice the shot group getting progressively wider.

Barrel wear is first noticeable in the area immediately in front of the chamber, known as the throat of the barrel. Generally speaking, higher-velocity cartridges wear barrels faster than slower cartridges.

Depending on the cartridge, the load used to propel the bullet can last from three thousand rounds in an ultra-high-velocity wildcat cartridge to ten thousand rounds in a .308 barrel. Because of this variability, it is important to keep an accurate log of the rounds fired through the rifle's barrel.

Twist

The addition of rifling to barrels first occurred in the late 1400s and early 1500s. Like the muskets of old, shotguns do not have rifling and are referred to as smoothbore.

The rifle's helical grooves cause the bullet to turn on its axis as it travels down the barrel and then spin at a high rate of speed as it exits the barrel on the way to its target. This spinning motion greatly enhances the bullet's stability, just as a perfect spiral motion enhances the accuracy and flight of a well-thrown football.

Most barrels have a small section, immediately after the chamber, that has no rifling. It is known as the free bore or jump. This jump length is very short and in some custom rifles can be eliminated; more about this in chapter 6.

The amount of twist (i.e., the number of rotations per inch of barrel) is important for the accuracy of the bullet. Small differences in twist rate can have major accuracy influence. A good rifle barrel, with the right twist and a high-quality bullet, can perform magically.

There are many theories on what exact twist is best. Today, most people acknowledge that there is an ideal twist for each caliber, bullet, and barrel length, but when in doubt, it's always better to go for the higher twist rate.

Selecting the right twist rate requires taking multiple factors into consideration, including: a) a bullet that is longer in proportion to its diameter will require a faster twist to stabilize it; b) long-nosed bullets, such as hollow-point boat tails, have less drag and increased aerodynamics, but are harder to stabilize; c) longer and heavier bullets can be stabilized with a twist suitable for a lighter bullet if you increase the bullet's velocity; and d) a slower bullet requires a faster twist.

≈ Another bullet's-eye view of a rifle barrel's twisting pathway

For hunting and all-purpose rifles, it's best to choose the twist rate that will work satisfactorily with the biggest and heaviest bullets made for that rifle. For precision work, on the other hand, the ideal twist is one that best matches and meets the needs of the specific bullet chosen for the work at hand.

Fortunately, there are smart engineers and gunsmiths who work this out for us, so we don't have to lose sleep at night wondering if our twist ratio is appropriate for our ammo.

More points to know about twist:

1. If the rifling lands extend to the end of the chamber, less gas escapes and pressure is increased. This permits a lower powder load for the same velocity.

2. Long, pointed bullets are harder to stabilize because any error in manufacturing will cause problems with the bullet's center of gravity and affect its gyroscopic stability in flight.

3. Plain, short bullets work best with a slower twist.

4. If two bullets are fired simultaneously with the same velocity and spin rate, the heavier, longer bullet will maintain its speed of rotation better than the shorter, lighter bullet.

5. The best rifle twist is governed by the bullet's length, not its weight.

≈ SEAL sniper Matt Johnson gets some trigger time in; probably not too concerned about rifle twist right now. (*Matt Johnson*)

6. Barrel manufacturers have tolerances for the actual twist, e.g., a 1-in-12-inch twist may actually range from a 1-in-11-inch to a 1-in-13-inch twist. Custom barrels can be made more precisely but at an increased cost.

7. Increasing twist for a faster rotation speed will not give a flatter trajectory.

8. Rechambered rifles commonly do not shoot as expected because the twist is not correct for the new caliber.

9. If the twist is not correct for a bullet, switching to a heavier, longer bullet will not improve accuracy because the twist is too slow.

From Trigger Pull to Bullet Flight

As noted above, a tremendous amount of science goes into building an accurate sniper rifle. Becoming a top sniper does not require that one knows or understands all the nuances of the sciences of metallurgy, propellant (powder) chemistry, and ballistics. However, the more a shooter understands the science, the better he or she will understand why certain shooting facts make sense, and be able to use these facts to become a better sniper.

There is also a tremendous amount of science involved in what happens to the sniper's rifle once the trigger is squeezed, and it is just as valuable to have a decent grasp of that science, too.

The phenomena described below all happen in the microseconds following the firing pin striking the primer, so it's nearly impossible to actually see these events as they happen, but it's not hard to conceptualize them.

Once the firing pin strikes the primer, the primer compound explodes and ignites the powder in the cartridge case. The resulting hot gas expands in all directions—not just against the bullet's base and the rear of the rifle chamber. The pressure

⌃ **Becoming a master at the craft requires range time. A SEAL sniper's three-round shot group at three hundred meters on a steel plate with his .300 Win Mag rifle.**

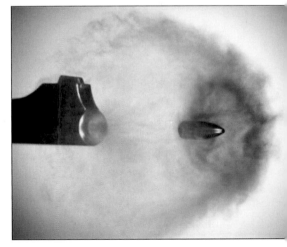

⌃ **Internal-to-external bullet transition (*Andrew Davidhazy*)**

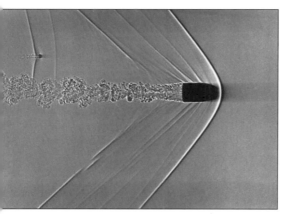

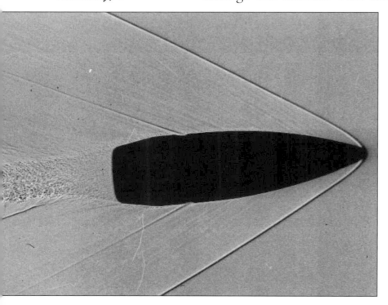

created causes the barrel to expand, and the amount of this expansion or swelling depends on the type of barrel, its thickness, and the type of steel, as well as on the amount of powder ignited.

This expansion moves down the barrel and decreases as the pressure drops during the bullet's travel. Thus, it is greatest at the breech end of the barrel.

Picture a sausage-shaped balloon being squeezed from one end to the other: The balloon expands just ahead of the pressure. The same things happens to a rifle barrel. Experiments have shown that if you fire a bullet into a water tank, the bullet diameter will be larger as you cut the barrel shorter because the barrel expands more at the breech end. It makes sense that heavier, thicker barrels will not expand as much as thinner barrels.

In addition to expanding in all directions, the barrel is also twisted violently by the bullet being forced to rotate by the barrel's rifling. The force or torque required to start the bullet rotating causes a distortion on the barrel that increases as the bullet moves from the action through the muzzle. This rotation is counter to the rotation of the bullet and precedes the bullet down the barrel. These two phenomena explain the importance of having a free floated barrel.

Not allowing the barrel to come into contact with anything as the rifle is fired allows the barrel to move unimpeded, in the way its manufacturer intended, and provides for optimal accuracy.

These forces also place tremendous stress on the steel, in addition to the heat generated by the friction of the bullet and the hot gasses in the barrel. With enough stress, cracking and eventually barrel failure can occur. Any weak points in the grain texture of the steel are the most common places for these cracks to occur, which highlights the importance of a barrel being stress relieved during the manufacturing process.

It's natural to assume that a rifle barrel would be perfectly straight, and that the strength of the steel would prevent the barrel from drooping or sagging. Wrong. Any material supported on only one end will sag slightly.

Here is where this gets truly fascinating: Yes, the barrel will inevitably droop, albeit so slightly as to be imperceptible—but upon firing, as the bullet travels down the bore, the forces exerted on the barrel will try to straighten it. These forces create shock waves that produce vibrations, or harmonics, and cause the barrel to move in a whiplike fashion, called barrel whip. As you sight the rifle in, your optics will correct for this motion—as long as the motion remains consistent.

This again points out the need for the barrel not to be hindered from moving as it does naturally with every shot. Even improper use of a sling attached to the barrel or the stock and

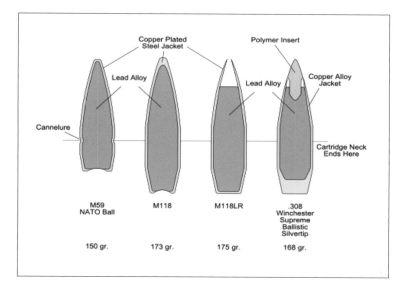

Copper Plated Steel Jacket

Polymer Insert

Lead Alloy

Lead Alloy

Copper Alloy Jacket

Cannelure

Cartridge Neck Ends Here

M59 NATO Ball

M118

M118LR

.308 Winchester Supreme Ballistic Silvertip

150 gr.

173 gr.

175 gr.

168 gr.

« Cross-section of some popular rounds and their compositions; the two rounds on the left are full metal jacket, the two on the right are considered "open tip." Aerodynamically speaking, there isn't much difference between the two M118 rounds in the center.

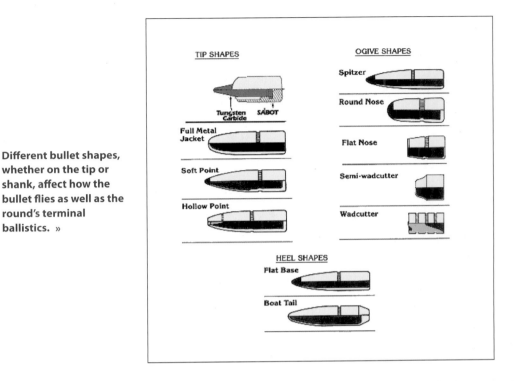

Different bullet shapes, whether on the tip or shank, affect how the bullet flies as well as the round's terminal ballistics. »

coming into contact with the barrel can alter the bullet's point of impact. Barrel whip will change with different ammunition, different powder loads, and different bullet weights, which reinforces the point: A high rate of accuracy is the result of a high rate of consistency.

As you can see, there is more to being an effective sniper than meets the eye. The amount of technical data and knowledge of the materials and physics involved that a modern sniper absorbs will directly relate to his lethality in real-world shooting situations.

Next time you throw a round off target while resting your barrel against a support, it will all become clear to you. Knowledge is power.

⌃ Gotcha. A SEAL conducting over-the-beach (OTB) operations. It's common for SEAL snipers to sometimes use machine guns to lay down accurate fire. A SEAL sniper on an M60 or Squad Automatic Weapon (SAW) is a deadly combo.

6

WHAT MAKES A GOOD SNIPER RIFLE?

Rifles have evolved over time in design and manufacture to be more effective at hitting and killing the intended target, whether game or humans.

Now, it doesn't take any special large, high-velocity round to send a human being into taking a dirt nap. A well-placed .22-caliber air rifle round is perfectly capable of killing a human being. (Robert Kennedy was gunned down with a .22-caliber handgun.)

So what characteristic sets a sniper rifle apart from your typical hunting rifle? One word: *accuracy*.

Sniper-grade rifles must be capable of superb accuracy under a wide range of shooting conditions. There is no room in a sniper's toolbox for an inaccurate rifle. Lives, including one's own but often including many others, may well depend on that weapon doing its job.

Degrees of Accuracy

⌃ **The Accuracy International's AS50 semiauto .50 BMG; the acronym designates this as a Browning machine gun .50-cal round, but it's also known as the "Badass Mackdaddy Gun."** (*Accuracy International*)

The spread created by MOA (*Reeds Target Shooting Club*) ⌄

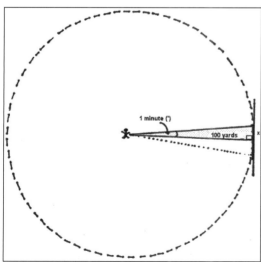

1 minute (')

100 yards

We've mentioned the term *minute of angle* (MOA) before, but it's time to define the term, because MOA is a critical acronym in the sniper's vocabulary. To a sniper, a rifle's potential MOA refers to how accurately it is capable of shooting out to different distances.

Minute of angle is a measurement of arc, or angular width. A complete circle is divided into 360 degrees, each of which is divided further into sixty *minutes*. Therefore, one minute is one-sixtieth of a degree. (Each minute is also subdivided into sixty *seconds*, but that doesn't concern us here.)

At one hundred yards, 1 minute of angle creates a spread of 1.0472 inches; in most shooting literature, 1 MOA is usually rounded as 1 inch at one hundred yards. The spread created by MOA increases linearly as you go further from the muzzle, i.e., 2 inches at two hundred yards, 3 inches at three hundred yards, and 10 inches at one thousand yards.

Rifles, especially precision rifles, are advertised to be capable of shooting a certain MOA, the lower the number the better. Most commercial hunting rifles shoot 2 MOA or greater.

A 2-MOA hunting rifle would at best be capable of shooting a three-to-five-shot group into a two-inch circle at one hundred yards, a four-inch circle at two hundred yards, or a ten-inch circle at five hundred yards. If you are shooting an elk, which has a kill zone of twenty-four inches, this accuracy is acceptable, as long as you can center the shot somewhere within the kill zone.

Imagine you are called upon to take an eight-hundred-yard shot with that same 2-MOA rifle. At eight hundred yards, 2 MOA translates into *sixteen* inches. The average male torso at the shoulders is twenty inches, tapering to twelve inches mid-chest. The average human head measures six to seven inches in width, and eight to ten inches in height. With those target dimensions, a sixteen-inch spread doesn't give you much margin of error!

Obviously, making an accurate shot with this rifle would require a lot of skill—and a fair amount of luck. Suddenly the rifle that served you well while hunting deer could easily cause you to miss your target, or worse, take out a friendly.

Okay, so a sniper rifle has to be accurate—but how accurate is accurate enough?

As noted above, most hunting rifles are usually good for 2-to 3-MOA accuracy. Sniper rifles, on the other hand, are built to have a potential accuracy of *at least* 1 MOA, and less than 1 MOA (called sub-MOA) is preferred, with most sniper rifles being .5-MOA-capable. To do the math: a .5-MOA sniper rifle will be capable of shooting a .5-inch group at one hundred yards, or a five-inch group at one thousand yards. Now that's tight!

As with most things in life, you get what you pay for. A 1-MOA rifle will probably be less expensive than a more accurate rifle because less expensive parts were used, it is built to looser tolerances, and it has less hand-built attention in its manufacture.

The Remington Police Sniper Model 700 .308 rifle is a good example of a 1-MOA rifle. This rifle will cost about $1,000. With today's materials and knowledge, sniper rifles can be built to be capable of at least .5 MOA. The Remington M24 HS Precision Tactical rifles are good examples of .5-MOA sniper rifles. The cost of this grade of rifle is about $2,000 to $2,500.

Some rifles, given the right ammunition and right caliber, are capable of .375 to .25 MOA and are accordingly that much more expensive. Sniper rifles from GA Precision, Tactical Operations,

⌃ **The Knights Armament M110 SASS (semi-automatic sniper system), a 7.62 match-grade system with a lot of adaptability, is shown here with the digital desert camouflaged UNS LRLP (universal night-sight long-range low profile), otherwise known as the PVS-26—an amazing setup for night operations. (*Knights Armament*)**

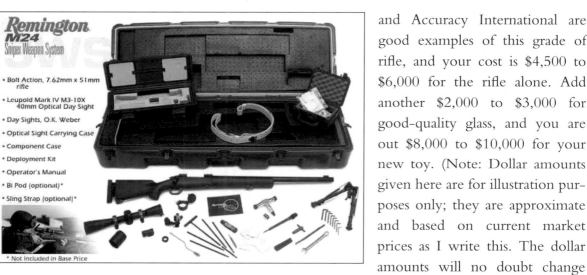

Remington
M24
Sniper Weapon System

- Bolt Action, 7.62mm x 51mm rifle
- Leupold Mark IV M3-10X 40mm Optical Day Sight
- Day Sights, O.K. Weber
- Optical Sight Carrying Case
- Component Case
- Deployment Kit
- Operator's Manual
- Bi Pod (optional)*
- Sling Strap (optional)*

* Not Included in Base Price

⤒ **The authors would love to find this complete kit under the Christmas tree some year. (*Remington*)**

and Accuracy International are good examples of this grade of rifle, and your cost is $4,500 to $6,000 for the rifle alone. Add another $2,000 to $3,000 for good-quality glass, and you are out $8,000 to $10,000 for your new toy. (Note: Dollar amounts given here are for illustration purposes only; they are approximate and based on current market prices as I write this. The dollar amounts will no doubt change over time.)

Let's look at some grade components.

The Barrel

This Barrett Model 95 is a shorter, lighter version of their bolt-action .50 cal, the Model 99, and sports a (relatively, for a .50) short twenty-nine-inch barrel. (*Barrett*) ⤓

As I mentioned, sniper-grade barrels are constructed from either chromoly or stainless steel. Stainless is more expensive and may resist rust and wear better than chromoly. Each manufacturer will, of course, claim that their barrel is the best and tell you why their way of manufacturing results in a high degree of accuracy. Again, you get what you pay for.

In general, most sniper-grade barrels are thicker than commercial hunting rifle barrels. The extra thickness is beneficial to the barrel's accuracy because it is stiffer and therefore has less whip and motion during bullet transit. The additional thickness also makes the rifle heavier, especially toward the front of the rifle. This makes the shooter feel less recoil, again adding to his accuracy.

The added forward weight also decreases muzzle jump, which helps the shooter stay on target and rapidly put another accurate round down-range if necessary. The added thickness also helps deal with the heat of repeated fire.

The added weight, of course, is also a concern to the sniper having to carry the rifle into the field, but it does translate into improved accuracy and in most cases is an acceptable trade-off.

Besides being thicker, most sniper-grade barrels are also longer than hunting rifles. The barrel's length should be tailored to the caliber and the powder load.

Again, you want the barrel to be long enough to maximize the full benefit of the gas developed by the powder. The shorter the barrel, the less the muzzle velocity; too long a barrel with an improper powder load could also reduce pressure, which could result in a lower muzzle velocity.

Shorter barrels are stiffer; however, accuracy may suffer, especially at longer ranges, again because of the decreased velocity. Shorter barrels are easier to maneuver, but most snipers will not be doing house-clearing with their rifle, and shooting offhand or in nonsupported positions is easier with a shorter, lighter rifle.

Most sniper rifles, especially for 720- to 900-plus-meter shooting, will have 60- to 67.5-centimeter barrels. Additionally, a shorter barrel will have a louder muzzle blast (due to an increased density of gas being released), which, if unsuppressed, could be a concern to a sniper trying to hide his location from the enemy.

Another important factor in a quality barrel is how it is set in relation to the stock. As I mentioned in chapter 5, the barrel should not come into contact with anything when the rifle is fired, including the stock. A quality sniper rifle will have the stock and the rifle barrel set so that the barrel does not come into contact with any part of the stock; this is known as "free-floating the barrel."

You can check to see if the barrel is free-floated by running a sheet of paper folded in half along the barrel above the fore end of the stock. Some .50-caliber builders will eliminate this problem by not having a fore end to the stock. This principle now appears in a lot of AR-style rifles where the fore end is a free-floated tube that does not touch the barrel at all.

Test for a free-floating barrel (www.tikka.fi)

The Rifling Process

Rifling varies from manufacturer to manufacturer. As I mentioned in chapter 5, the cheapest method is hammer forging. Some very accurate rifles—the Remington 40 xb, the Remington M24, and Sig Sauer rifles—use this method to cut the barrel rifling. Custom barrel manufacturers, such as Bartlien, Schneider, Lawton, and Krieger, employ cut rifling exclusively.

Modern technology tolerances for the bore and the rifling can be as small as .0005 centimeters, which, as those of us who managed to stay awake in fourth- grade math know, is very tight.

All the various rifling processes leave behind at least some tiny imperfections that then need to be removed with some kind of lapping process. Lapping will make a mediocre barrel better but not great; what really matters is quality throughout the entire barrel-building process.

You might wonder why it is important to remove theses imperfections from the barrel. After all, the bullet is being pushed down the barrel typically at two to three times the speed of sound.

How could such small imperfections in the barrel degrade its accuracy?

The answer is that these imperfections, no matter how small, interfere with the bullet's moving smoothly down the barrel. They also strip away parts of the bullet's copper jacket, so the bullet leaves the barrel not perfectly symmetrical, which interferes with its gyroscopic stability during its flight to the target. The copper left behind in the barrel can build up, adding to the roughness of the barrel's bore and the fouling of the barrel during subsequent shots. An unstable bullet is an inaccurate bullet.

Most manufacturers will recommend some kind of break-in process designed to smooth out these imperfections. The break-in process is time-consuming, usually requiring a series of shots, cleaning the barrel between single shots, double shots, and finally three-to-five-shot groups. The theory behind this is that as you alternate shooting rounds and cleaning the barrel, these small imperfections are shot smooth.

There is also a line of abrasive bullets available from David Tubb, the legendary long-distance rifle champion, that are designed to smooth the barrel more quickly and more efficiently than the laborious break-in period.

Still, if the barrel manufacturer recommends a certain break-in process, it may be best to follow their routine so that they will

⌃ **Major league baseball player Kevin Kouzmanoff getting some range time with Brandon's .300 Win Mag. Notice the rear accessory pack and speed loader.**

Threads for muzzle brake or suppressor

Crown-note bevel

Bore

Land and Groove Of Rifling

Tactical Operations Muzzle Threaded for a Muzzle Brake or Suppressor

⌃ **Detail: the pointy end of the spear (*Curtis Prejean*)**

honor any warranty. There are also products on the market that can be applied to the barrel to coat the bore. Again, the barrel or rifle manufacturer may recommend these products, based on their experience and their product.

Twist, Crown, and Jump

The twist rate is the rate at which the helical or spiral pattern turns over a certain length of barrel. For example, a twist rate of 1 to 10 means that the rifling completes one complete revolution every ten inches.

Again, it is this twist that causes the bullet to rotate as it moves down the barrel, giving the bullet gyroscopic stability during flight. Having the proper twist is very important to accuracy; the twist is what causes the bullet to rotate as it moves down the barrel and imparts its gyroscopic stability in flight, and an unstable bullet is an inaccurate bullet.

The proper twist rate can be determined with ballistic formulas that are too detailed to go into here. (If you are interested in the math and science of this subject, check out *Understanding Firearm Ballistics*, by Robert A. Rinker.) Rifle barrel manufacturers will usually match the twist rate to the intended ammunition for the sniper rifle.

Most sniper-grade .308s, .300 Win Mag, and .338 Lapua rifles have a 1-to-10 twist. As a general rule, the heavier bullet will require a faster twist, so if you know that you will only be shooting a 300-grain .338 bullet out of your .338 Lapua sniper rifle, you may want a 1-to-9 or 1-to-5 twist rate. Again, custom-built rifles can have this kind of tweaking, but it will cost more money.

The crown on a rifle barrel refers to a bevel cut at the muzzle. The crown is the last part of the rifle that will come in contact with the bullet as it leaves the rifle. Any

≈ **SEAL heads north from Baghdad manning the .50.**

imperfections, indentations, or nicks in the bore at the end of the barrel will impart some instability to the bullet.

Quality sniper-rifle manufacturers will bevel the end of the barrel to ensure that the bore is free from any imperfections. Usually this machining is beveled in so the crown is recessed from the end of the barrel to help protect the crown or bore from being hit or dinged during operations. If the barrel is threaded to take a suppressor, there is usually a crown protector that can be put on the end of the barrel.

If the crown becomes damaged, you will see a drop in accuracy. Fortunately, a quality gunsmith can recrown a damaged barrel, so don't despair if you drop yours muzzle-first off the top of a ten-story building.

The jump, also called bullet jump, refers to the distance from the end of the chamber to the beginning of the rifling.

As noted earlier, rifle manufacturers typically cut the chamber to match standard cartridge dimensions. Most sniper rifles will have this as the standard chamber. Why? Because the rifle has to shoot all brands of ammunition issued to the sniper. There will be slight deviations in cartridge length from manufacturer to manufacturer and from lot to lot. (A lot is the quantity of ammunition assembled by one producer under similar conditions and is expected to produce similar results on target.)

Most snipers, especially in the military, do not have the option of hand-loading all their rounds; they shoot what they're given, so their rifles must be capable of shooting all ammo.

If you can hand-load and you want to customize your rifle further, you can have the barrel-builder shorten the jump to tighter tolerances, or seat your bullets out slightly longer during the reloading process.

Some reloaders try to have the ogive (curved sides) of the bullet actually touch the lands and grooves of the barrel. As always, there are trade-offs for doing this, and it may yield only a slight increase in the rifle's accuracy. Bench-rest shooters are more likely to benefit from this than are active snipers.

Also note that having your chamber cut to match the cartridge's overall length precisely, so that there is minimal to no bullet jump, is for lead core bullets. You should not do this if you are going to use solids, since solid bullets do not conform to the barrel as easily as lead core bullets. You could run into problems with very high pressures that could damage your rifle—or damage *you*.

Life Expectancy

No, not yours. The question here is, how long will a sniper-grade barrel last?

The answer depends on several factors. Stainless steel is a bit more wear-resistant than chromoly, but the three biggest factors affecting barrel life and long-term barrel accuracy are the velocity of the round, the number of rounds through the barrel, and barrel care and maintenance.

High-velocity rounds (greater than 3,500 to 4,000 fps) will wear a barrel much more quickly than slower cartridges. There is a trade-off here in terms of wanting a flatter trajectory, less wind interference, higher energy, as weighed against being willing to buy a new barrel sooner.

The number of rounds through the barrel is important to keep track of because over time you will see a decrease in accuracy as your barrel wears from shooting. (See the sample Barrel Log in Appendix A.)

Maintenance of your barrel—good cleaning, not scratching the bore, and keeping the crown from getting damaged—will make the barrel last a long time. Want an estimate for a sniper-rifle barrel life? A good sniper-grade .308 barrel that's well maintained should normally last a good eight thousand to ten thousand rounds.

Triggers

As we said in chapter 3, there are two basic types of triggers: single-stage and two-stage.

Single-stage triggers are the most common type used in American-made sniper/tactical rifles. In a single-stage trigger, there is virtually no uptake in the movement of the trigger as the shooter applies pressure to break the shot. The pressure required

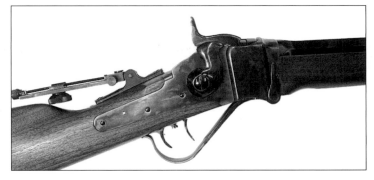

≈ Here's an old-school double-set trigger on a Sharps Model 1874 sporting rifle. The rear trigger is pulled first, which then allows the front trigger to fire the round. (*ammoland.com*)

to cause the trigger to "break" (fire) can be altered with some custom trigger work, and in some rifles the trigger is adjustable. You want the trigger pull to be smooth and consistent. A trigger job can smooth parts in the trigger assembly to improve trigger pull.

Most tactical/sniper rifles have a trigger pull of 2.5 to 3 pounds. You don't want a trigger that takes a tremendous amount of effort to break, but you also don't want a trigger that will break if you slam the bolt closed or that will break with the slightest pressure, before you are sure of your target and have the green light to engage the target. More weight will cause additional stress to the sniper during the natural respiratory pause generally recommended as the ideal time to break your shot.

A trigger can be adjusted to less than 2.5 pounds (all the way down to 8 ounces), but too light a trigger can result in accidental discharges. Remember, once you squeeze the trigger there's no calling the bullet back!

Two-stage triggers are so called because the trigger mechanism is engaged in two stages. The first stage involves pulling up the slack in the trigger movement, which is fairly light. In the second stage, the required trigger pressure is higher but remains constant until the trigger breaks.

Two-stage triggers are standard on Accuracy International rifles and other European rifles. It should be noted that most rifle actions will only work with one style of trigger; in most cases, you cannot switch out a single-stage trigger assembly for a two-stage trigger, even if the two-stage trigger is what you prefer.

Remington triggers are easily tuned and Remington offers adjustable triggers on some of their higher-end rifles. After-market triggers from Timmey and Jewel are available for rifles built with a Remington Model 700 action (Tactical Operations and HS Precision).

⌃ The J. Allen Enterprises JAE-700 RSA Precision rifle stock for the Remington 700 short action is part of a new generation of fantastic stocks on the market. (www.jallenenterprises .com)

Sniper Rifle Stocks

"This is my rifle. There are many others like it, but this one is mine. My rifle is my best friend. It is my life. I must master it as I must master my life. Without me, my rifle is useless. Without my rifle, I am useless. I must fire my rifle true. I must shoot straighter than my enemy, who is trying to kill me. I must shoot him before he shoots me. . . ."

This part of the Marine Corp's "Rifleman's Creed," made famous by the 1987 movie *Full Metal Jacket*, certainly applies to all snipers.

It is the stock that lets a shooter hold, nestle, and fit a rifle to his body, becoming one with the rifle. To accurately place rounds downrange, the sniper must be able to consistently position the rifle with his body and support it the same way on every shot. Put another way, a sniper needs a rifle that "hits and fits." In order to accomplish this, most sniper-grade rifles have adjustable stocks that can be changed to fit the sniper. Like a finely tailored suit, the rifle needs to be fitted to the shooter.

The two parts of the rifle stock that are adjustable are known as the length of pull and the cheek well. The length of pull is the

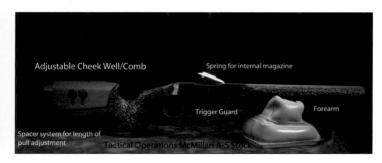

⌃ Tactical Ops stock from McMillan; nice (*Curtis Prejean, courtesy of McMillan*)

distance from the place you put your firing hand to the butt of the rifle stock. A quality sniper-rifle stock can be lengthened or short-ened as needed by the sniper.

Why would this distance need to be changed? Shooting with full body armor, as opposed to shooting in a plain uniform, will add size to the sniper. Shooting in the dead of winter will usually have the shooter dressed in warmer, bulkier clothes as opposed to the dog days of summer. Additionally, different shooting positions will change the distance the shooter needs to get proper eye relief with the optics.

The cheek well is the top edge of the rear portion of the stock, where the sniper rests his cheek during shooting. This, too, needs to be adjustable, depending on the shooter's head and facial structure and the height of the optics on the rifle.

The height of the cheek well should be adjustable so that the sniper can place his cheek on the rifle in the same place every time. It should be adjusted in such a way that does not strain his neck or head or cause him to contort himself to the rifle—so that the sniper could practically fall asleep on the stock.

Despite what Hollywood shows, not all sniper missions entail shooting hundreds of rounds. Often there are long periods of observation, calling in air strikes or waiting for the target to pre-sent a shot. The rifle must fit the sniper, not the other way around, so that he can comfortably position himself and his rifle to observe the enemy for long periods of time.

Detailed view of an HS .308 Witra McCree aluminum stock (*HS and McCree*) ⊻

Today most sniper-grade rifles are made of synthetic/fiber-glass materials rather than wood. The fiberglass stocks from com-panies such as McMillan, HS Precision, JAE, and Mannard are very strong and rigid and do well in a wide range of demanding operating condi-tions. Many of these synthetic/fiberglass stocks have aluminum

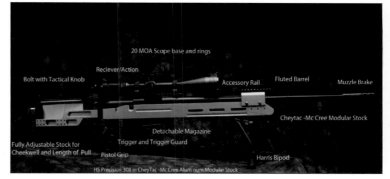

20 MOA Scope base and rings
Reciever/Action
Bolt with Tactical Knob
Accessory Rail
Fluted Barrel
Muzzle Brake
Cheytac -Mc Cree Modular Stock
Detachable Magazine
Fully Adjustable Stock for Cheekwell and Length of Pull
Trigger and Trigger Guard
Pistol Grip
Harris Bipod
HS Precision 308 in CheyTac -Mc Cree Alum num Modular Stock

pillars for holding the action in the stock, but the action should be bedded into the stock with additional epoxy or fiberglass so that the action is snugly nestled into the stock.

There is a trend toward using an all-metal or modular stock, or chassis, that does not require any bedding. Accuracy International rifles have an aluminum chassis that simply requires that the rifle action be secured to the chassis with the appropriate screws with the proper amount of torque. McCree and the new Remington MSR sniper rifle use modular aluminum stocks.

Sniper stocks, unlike the stocks on hunting rifles, typically feature areas on the stock forearm that are designed to hold a variety of sniper accessories, such as lights, lasers, night-vision optics, and bipods.

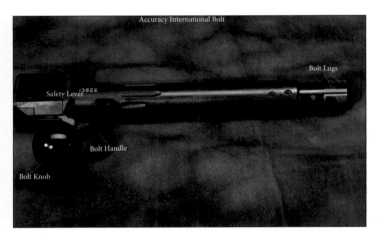

≈ **Accuracy International bolt (*Accuracy International*)**

The Remington Mil-stock R bolt and action in an Accuracy International aluminum chassis; note the machine work. (*Accuracy International*) ≈

The Bolt and the Action

The bolt and the action should be machined so that the bolt's locking lugs match the action and are precisely aligned with the chamber and bore of the rifle. The bolt face that comes in contact with the base of the cartridge should be perfectly flat so that the cartridge cannot be bent, warped, or twisted in the chamber. Any misalignment will result in decreased accuracy.

The firing pin housed in the bolt should be set to strike the cartridge with enough force to detonate the primer, but not so much as to cause excess vibration in the rifle.

"Lock time" refers to the time that elapses between the moment your trigger finger releases the trigger mechanism and when the firing pin strikes the primer. Lock time should be as fast as possible so that the rifle has virtually no time to move between the instant you actually activate the firing pin and the resulting ignition of the powder. Lock times vary from 0.0022 to 0.0057 second. You can have the lock time adjusted with different springs and also by replacing the steel firing pin with one made of titanium. Again, you get what you pay for.

⌃ **McMillan 3A stock on a Fulton Armory M14 rifle** (***Gun Tests***)

If you hear that a bolt has a "tactical" knob, this simply means that the bolt operation lever is a little longer and may have a bigger knob on it, compared to a nontactical rifle, since a sniper may be using gloves. However, the biggest reason for a tactical bolt lever or knob is that under the stress of combat, your fine motor skills decline at least 40 percent; having a larger knob to grab and manipulate may be a benefit when you are being shot at.

Muzzle Brakes

Depending on the caliber or the shooter's preference, the barrel may be fitted with a muzzle brake. A muzzle brake will definitely reduce the felt recoil, but it will also increase the sound of the rifle firing. Most muzzle brakes will, on average, reduce recoil about 30 percent.

Smaller-caliber rifles (.223, 6.5 mm, .308 Win) may not be fitted with a muzzle brake, since the recoil of those calibers is not an issue for most shooters. However, the muzzle brake's reduction in recoil and barrel jump do help keep the rifle on target and allow the sniper to put additional shots downrange more quickly.

For larger cartridges (.300 Win Mag, .338 Lapua, .408 CheyTac, and .50 BMG), muzzle brakes are much more common because without the brake the recoil would be much more difficult to manage.

To some snipers a suppressor is the ultimate muzzle brake. Suppressors both reduce felt recoil *and* drastically reduce the sound of the rifle firing—a tremendous benefit for a sniper in hiding from the enemy.

If you plan to shoot with a suppressor, you should be sure to zero the rifle both with and without the suppressor, as the addition of a suppressor will to some degree change the weapon's zero. We'll look at suppressors in chapter 8.

Summary: What Makes a Great Sniper Rifle

1. A match-grade, free-floated barrel. This will be thicker than regular hunting rifle barrels and manufactured with tighter tolerances. Ideally this will give the barrel the potential for sub-MOA accuracy. The barrel should have been stress-relieved during manufacturing and should be lapped to reduce as many of the defects associated with boring and rifling process as possible. *Break in your barrel according to manufacturers' specifications.* If you don't, you'll sacrifice barrel life and accuracy.

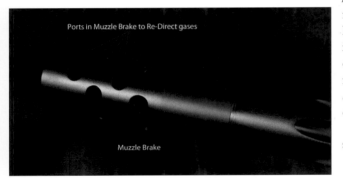

Ports in Muzzle Brake to Re-Direct gases

Muzzle Brake

⌃ **One of the many different styles of muzzle brakes** (*Accuracy International*)

2. An action mated to the weapon constructed with the highest standards.

3. If shooting a large-caliber weapon, ensure you are using a muzzle brake or suppressor. Your shoulder will appreciate it if you're putting a lot of rounds downrange, and if you are shooting from a concealed position, the suppressor will ensure you *stay* concealed.

4. An adjustable stock, quality bipod, and the best optics you can afford.

« **This Remington 700 has a Speedlock titanium firing pin kit with the U.S. Optics SN3 variable-power scope and a Harris bipod, and was built by ARM USA's Darkhorse Gunworks.**

« Glen getting the feel for Springfield Armory's rerelease of the M21 sniper system, basically an M1A Supermatch with a custom Douglas barrel and an adjustable cheek piece stock. Brandon loves this rifle.

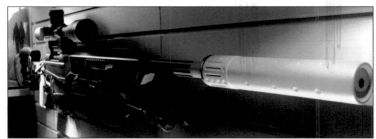

« Ashbury International's Asymmetric Warrior ASW338LM Precision Sniper Rifle System. This weapon uses Rock Creek Pinnacle barrels and comes equipped with an integral muzzle brake and sound suppressor system. This cutting-edge U.S. company is pushing the technological envelope with their weapons systems. Expect to see more from them.

« Example of a sniper-grade rifle: the Accuracy International .338 Lapua Magnum (*Curtis Prejean*)

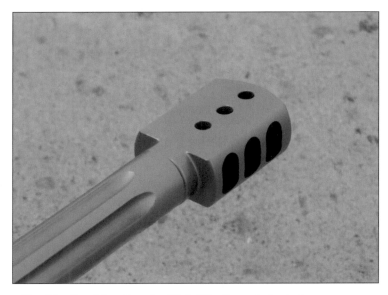

⚹ The Allen Precision Shooting "Painkiller" muzzle brake (*www
.longrangehunting.com/forums/f19/new-slim-painkiller-muzzle-brake
-aps-33319/*)

Accuracy International muzzle brake (*Curtis Prejean, courtesy of Accuracy
International*) ⚹

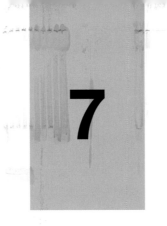

7

BEGINNER BALLISTICS, EFFECTIVE RANGE, AND POTENTIAL APPLICATIONS

Humans do not do well with bullet holes in them. Granted, there are some instances in which a human being can sustain multiple gunshot wounds and continue to function for a time, depending on the extent of the injury, the location of the injury, and access to appropriate medical care.

It's true, there are many examples of a drug- or adrenaline-fueled combatant requiring multiple rounds to stop him. In fact, it is well known that even with a shot to the heart, an appropriately motivated enemy can live on for as much as five to eight seconds—more than enough time to shoot you in a close-quarters gun battle.

But still, by and large, you hit a guy hard with a little chunk of metal traveling as near (or beyond) the speed of sound, and he's probably going down.

Given the fact of human frailty, why worry about what caliber you use? Why not use just any caliber?

Two main reasons: To accomplish his mission, the sniper needs a) the right combination of rifle and caliber and b) an accurate caliber/bullet combination to match his sniper rifle.

Accurate caliber, you say? Yes: Not all calibers and bullets are created equal.

There are three main types of ballistics: internal, external, and terminal.

Internal ballistics relates to what is happening in the rifle as the bullet leaves the neck of the cartridge. *External ballistics* describes what happens to the bullet as it flies toward the intended target. *Terminal ballistics* is what happens to the bullet when it hits the target. Sniper cartridges, like sniper rifles, must be accurate out to the full range to which the sniper has to deliver the shot. Furthermore, the cartridge/bullet combination must have the proper terminal ballistics to ensure that the bullet has the velocity and energy to accomplish its mission on impact.

Different targets require different bullets. A bullet's ability to take out a target depends on the energy or force needed to penetrate and destroy the target. A head shot at 540 meters will require a certain amount of bullet energy to penetrate the skull. If the enemy is behind a glass windshield, a door, or a cinder-block wall, obviously the bullet will need more energy to penetrate the cover and still produce a kill.

⌃ As of 2010, Knights Armament M110 replaced Remington's M24 as the army's primary sniper system; it is shown here with the PV S26 universal night sight short range (UNS SR). (*Knights Armament*)

Over the years, the military has done intensive studies on which caliber bullet is best for particular sniper operations. Six calibers are most frequently used in our current environment: .223 Remington, .308 Winchester, .300 Winchester Magnum, .338 Lapua, .408 CheyTac, and the .50 BMG. All these cartridges are very accurate, yet all have their advantages and disadvantages and will to some degree be chosen specifically depending on the sniper mission. (Other, less-popular calibers include the 7mm Magnum and the .419 Barrett.)

Out of these six calibers, the .308 Winchester and .300 Win Mag are probably the most common. They are plentiful, relatively inexpensive, useful in a wide range of missions, and very accurate out to their maximum effective ranges.

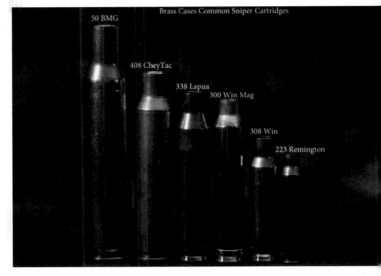

⌃ Calibers commonly utilized for sniper rifles (*Curtis Prejean*)

The .338 Lapua Magnum is seeing more use these days but is typically found only in the hands of a Special Operations unit. The .338 ammo is also much more expensive (a box of twenty costs $120 on the civilian market).

The .408 CheyTac and the .50 BMG both do extremely well at long rangers, that is, a mile to a mile and half—sixteen hundred to twenty-four hundred meters. They are also used for taking out hard targets, such as vehicles, radar, communication equipment, or engines. Several snipers from multiple coalition countries have set new records in the early years of the twenty-first century with confirmed kills at ranges beyond eighteen hundred meters.

Ballistic Coefficient

The exbal ballistic (external ballistics) charts in Appendix A show information for the 168-grain Sierra MatchKing bullet and the 167-grain Lapua Scenar bullet. Both are hollow-point boat-tail (HPBT) in design, but the Lapua Scenar has a better ballistic coefficient (BC), which is a measure of how well a bullet can overcome air resistance and keep its flight speed. The larger the BC, the better the flight characteristics. A perfect BC is 1.

The bullet's ability to keep as much of its muzzle velocity as possible during its flight is an important factor in both the trajectory and the killing effectiveness of the bullet. The characteristics that affect the bullet's BC are its mass, diameter, shape, and drag coefficient.

If you refer to Appendix C and compare the MatchKing and Scenar ballistic charts, you will note that the Scenar bullet has a slightly better BC: 0.47 as compared to 0.458. A very slight margin, perhaps—but note the difference in velocities, retained energies, and bullet drop!

Yes, a bigger BC is definitely better.

The .223 Remington

The .223 Remington was originally designed in 1957 for use in the military's AR 15/16. It is very accurate and relatively flat, shooting out to about 540 to 720 meters. At about 720 meters, the 77-grain bullet will go subsonic and become unstable. For sniper applications in which the target is a soft target, most consider about 540 meters as the effective range for this cartridge.

The .223 round does not have the necessary energy to penetrate ceramic plates in body armor. There have been instances in Iraq and Afghanistan where the .223 did not have sufficient energy to effectively eliminate a combatant who was high on drugs or had only minimal improvised armor.

Most sniper teams will have an AR-style weapon as a backup, for retreating if discovered, since the AR, with its semi- or full-auto operation and thirty-round magazines, offers a weapon that is easier to wield in closer combat, with easier target acquisition at ranges inside 360 meters.

You should be aware that unless otherwise designed and designated, bullets are supersonic when they leave the muzzle. There are special applications for subsonic ammunition. Ballistic science has shown that when bullets transit from

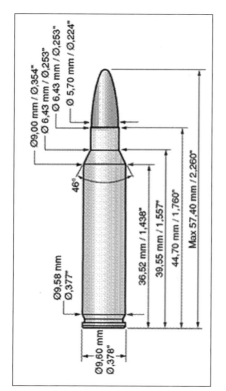

▲ **Exterior measurements of the .223 Remington**

« **Sound waves emanating from a discharging pistol**

supersonic to subsonic speed, they become unstable in flight and their accuracy greatly decreases.

The speed of sound will vary depending on a range of environmental factors. At sea level, in dry air at 68°F, it is 1,125 fps, or 343 meters per second (equivalent to 768 mph or 1,236 km per hour). When you consult a ballistic table or use a ballistic computer, note the distance at which the bullet drops below 1,125 fps. Beyond this distance, your chances of making an accurate hit on target become drastically reduced.

Again, in order for the bullet to take out the target, it will also need enough energy, which is dependent on its mass and the velocity at which it is flying. As anyone who went to sniper school (or took high school physics) will recall that kinetic energy is given by the formula $KE = \frac{1}{2} mv^2$, where "m" is the mass (not weight) and "v" is the velocity.

The .308 Winchester

The .308 Winchester was developed at the end of World War II, when the U.S. Ordnance Corps began looking for a smaller cartridge to replace the .30-06 Springfield. The cartridge went through numerous changes and was eventually adopted by NATO in 1953 as the 7.62 NATO. In 1957 the United States adopted it for use in the M14 and the M60 machine gun. Winchester introduced the cartridge to the public and chambered a Model 70 rifle for the cartridge, hence the name .308 Winchester.

The .308 Win is an extremely accurate round, thought by some to be the most inherently accurate .30-caliber cartridge ever produced. It is probably the most common cartridge for most sniper rifles, especially in missions that don't require shots past nine hundred meters or hard-target penetration at extended ranges. There is ceramic-plate body armor that will stop a .308, but Kevlar is not a problem for the round.

The most common bullet is the 168-grain hollow-point boat-tail, usually abbreviated as HPBT. The "hollow point" is not actually a true hollow point, as you would see in some pistol personal defense ammunition, but is a result of the copper jacketing process.

The .308 Win is a very accurate and versatile round. Most professionals would say that its effective range is about 720 meters, but in combat, snipers have successfully made shots on soft

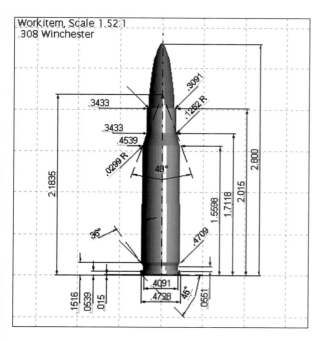

⌃ The .308 Winchester

Some 7.62mm (.308) Springfields on display at the SHOT show in Las Vegas »

targets out to 900 meters. In the exbal ballistic charts in Appendix C you'll note that the MatchKing .308 and Scenar .308 bullets remain supersonic at 1,000 yards, at 1,142 fps and 1,201 fps respectively. (The Scenar goes subsonic at just under 1,000 meters.) Even at this distance, both bullets retain energy that is equivalent to about one-third of a .44 Magnum at point-blank range.

The .300 Winchester Magnum (.300 Win Mag)

The .300 Winchester Magnum, usually called the .300 Win Mag, was introduced by Winchester in 1963. The .300 Win Mag is used in sniper rifles when there is a possibility of potential targets between 900 and 1,100 to 1,200 meters. This is more of a guideline than a rule, and certainly no reason not to shoot a close target with a .300. The .300 Win Mag cartridge for sniping is usually a 190-grain HPBT bullet.

This bullet goes subsonic somewhere between 1,200 and 1,250 meters. At this range the bullet still retains about the same energy as the .308 has at 900 meters. The .300 Win Mag is also a very accurate cartridge, but it does have more recoil than the .308 Win, and most .300 Win Mag sniper-rifles are therefore fitted with a muzzle brake.

The Remington M24 sniper rifle can be chambered for either the .308 or .300 Win Mag. All of the top sniper-rifle manufacturers have the .300 Win Mag as an option for those shooters requiring a rifle that is accurate out to the twelve-hundred-meter mark.

The .338 Lapua Magnum

The .338 Lapua Magnum was originally developed in the 1980s for the Navy SEALs as a long-range sniper round by Jerry Haskins of Research Armament Industries and Jim Bell and Boots Obermeyer. At that time, the cartridge was called the .338/.416, since its case was based on a .416 Rigby cartridge. Lapua then developed the .338/.416 further by strengthening the case and eliminating the original belted head design of the .416 Rigby. There-fore, the .338 Lapua is basically a necked-down .416 Rigby case that has been reinforced to withstand chamber pressures of 60,900 psi. (To use Bostonspeak, that's freakin' wicked high.)

The most common bullet weights in .338 Lapua sniper rifles are the 250-grain and 300-grain Sierra MatchKing HPBTs. Lapua also produces a 250-grain Scenar bullet that has a better BC than the Sierra MatchKings. The .338 is said to be effective out to 1,350–1,600 meters, although most shooters would stick with the 1,350- to 1,450-meter effective range figures.

If you feel like geeking out for a moment, check out the ballistic tables provided in Appendix C. A glance at these tables shows that the 250-grain MatchKing has a lower BC than the 250-grain Lapua. The MatchKing goes subsonic between 1,300 and 1,350 meters, while the 250-grain Scenar with a BC of 0.675 stays supersonic to 1,600 meters. Again, the bigger BC wins. The 300-grain MatchKing, with a BC of 0.768, despite its heavier weight and slower muzzle velocity, retains a supersonic velocity to 1,650 meters and even further! You can't escape physics.

The .338 Lapua is what most people would consider an intermediate sniper cartridge: heavier and more powerful than the .308 and .300 Win Mag, but not in the .408 CheyTac and .50 BMG league. The CheyTac has a muzzle energy of 7,700 foot-pounds and the .50 BMG an amazing muzzle energy of 12,000 foot-pounds, the .338 Lapua has 5,222 foot-pounds for the 300-grain Match-King. Because of this, the .338 can be used on some hard targets but does not have the penetrating power of the .408 or the .50.

The recoil from a .338 is very manageable, as long as the rifle is fitted with a muzzle brake. The .338 Lapua from Accuracy International is a joy to shoot and very accurate, with sub-.5 MOA at ninety meters. Shooting two-centimeter clay birds at nine hundred meters is no problem with this rifle.

⌃ **Brandon in the mountains of northern Iraq; good CheyTac country**

The CheyTac .408

If you are impressed with the potential capabilities of the .338, you better sit down, strap in, and hold on for the .408.

The story of CheyTac began a few years ago. Around the beginning of the twenty-first century, John D. Taylor saw that there was a need to improve the performance of the .50 BMG, which was developed in the early years of the twentieth century. (BMG is short for Browning machine gun. The cartridge was originally designed for use in a machine gun—not as a sniper cartridge.) Taylor brought together a team and created CheyTac Associates with a mission to develop and build a very long-range sniper cartridge.

The CheyTac team also saw a need for a cartridge that would fill the gap between the .338 Lapua and the .50 BMG, especially for soft-target (human) engagements, and they approached their mission in an unconventional manner: First they designed the ideal bullet, and then they designed the rifle to shoot the bullet.

Realizing that their bullet would have to possess outstanding ballistics at distance, they developed and patented a concept they called Balanced Flight.

The CheyTac bullets were developed by Warren Jensen, a partner and designer, at Lost River Ballistic Technologies, and were designed using PRODAS software and the Balanced Flight concepts. Jensen and associates designed a bullet where the linear drag is matched to the rotational drag, one of the features of Balanced Flight, to help the bullet retain its stability throughout its flight, giving it the ability to go farther and be more accurate.

The CheyTac designers also determined the ideal design of the barrel's lands and grooves, and found that the ratio of the bullet's total surface area to the total surface area of the lands and grooves should be somewhere between 3:1 and 4:1. Jensen experimented with a variety of calibers and found that improvements in bullet performance with the .408 were far beyond predicted.

CheyTac designers also looked into bullet composition. All the bullets we have talked about up to this point are composed of a lead core with a full copper jacket. The CheyTac bullets are made of a copper-nickel alloy, on a CNC machine (computer controlled) and lathe turned. These bullets are often referred to as solids, and the CheyTac's 305-grain bullet has a muzzle velocity of 3,300 fps at the muzzle.

When a bullet leaves the rifle muzzle, gravity immediately begins exerting its force on the bullet. In fact, it is interesting to note that if you were to have your rifle perfectly level on a bench at your local range, then simultaneously fire a round from the rifle and drop a second cartridge from the bench, the bullet leaving your gun and the cartridge dropped from the bench would hit the ground at the exact same moment.

THE LONGEST CONFIRMED KILL WITH A 7.62 RIFLE IN IRAQ

U.S. Army Sniper Reportedly Nails A Record 1250 Meter Shot With A 7.62mm Remington M24 SWS

◀ **Remington M24 SWS (Sniper Weapon System)**
Since its adoption by the US Army, the M24 is the standard by which all other military grade sniper rifles are judged. The M24 is world renowned as a long-range precision sniper system capable of enduring the harshest of military environments to include extreme high altitude and the depths of the ocean. The M24 is a combat proven force multiplier serving the US Army with honor and distinction against our nation's enemies; past, present and future.

⌃ **Remington struts its stuff.**

Because of this, even though it can appear to the casual observer that a sniper shoots a bullet in a straight line toward its target, there is no such thing as a perfectly straight bullet trajectory: Gravity is *always* a complicating factor.

Skeptical? The next time you're at the range, pay close attention to the angle of your rifle. It may look level, but the scope is dialing in elevation, gradually angling the rifle's aim upward to compensate for the effect of gravity as you increase the range to

≈ **The McMillan TAC-300 is your friend when you have a long way to travel.** (*McMillan*)

British snipers have easy access to Accuracy International's excellent rifles; all are made in Portsmouth, England. Shown here is the L115A3 .338 arctic warfare model. (*Lewis Page, The Register*) »

your target—because rather than following a straight line, your bullet will actually travel an arc-shaped path. At very long ranges it's almost as if you're lobbing your bullet into the target, the way you'd throw a baseball or softball.

The maximum height of the bullet's trajectory above the bore is known as the maximum ordinate, or max ord for short. Is knowing your bullet's maximum ordinate important? Yes or no, depending on the context. If you are always shooting in wide-open country, with nothing between you and your target except empty sky, then your max ord is irrelevant.

However, if you're shooting with any sort of overhead structure between you and your target, such as a bridge or tree cover, then knowing your max ord is important, even critical. You may think you have a clear shot; visually, that overhanging tree or bridge may appear to be completely above and out of your line of fire. But what your eye sees as your line of fire is not the path your bullet will actually follow. If that tree branch or bridge happens to be located at the height and distance where your bullet's trajectory reaches its max ord, the bullet will impact the structure—not your target.

Again, you may not even see the offending object in your scope, depending on the magnification, but it's there. Your carefully planned shot was missing one tiny detail—the bullet's max ord—and as a result, you were about to shoot the bridge, or worse yet, someone *on* the bridge, while your real target flipped you off and ran for cover, or returned fire.

≈ **Accuracy International's Super Magnum .338 Lapua**

The CheyTac 305-grain bullet is designed to have a max ord of only a few feet over the nine-hundred-meter distance, so a sniper could theoretically hold center mass on a human target from one to nine hundred meters and effectively engage his target. Not too shabby.

The CheyTac Intervention Rifle

The M-200 version of the CheyTac Intervention rifle is a bolt-action rifle with a seven-shot-capacity detachable magazine. The rifle and 419-grain bullet are proven out to 1,800-plus meters on soft-target, antipersonnel missions. CheyTac claims sub-MOA accuracy out to 2,700 meters in testing. Groups of 17.5 to 22.5 centimeters at 900 meters, 25 centimeters at 1,350 meters and 37.5 centimeters at 1,800 meters, have been obtained consistently. Personal experience has seen the rifle shoot a 12.5-centimeter group at 900 meters, and I've had first-round hits on a 30 x 50-centimeter plate at 2,000 meters.

⌃ **The CheyTac Intervention rifle (CheyTac)**

This extreme long-distance capability gives the sniper great standoff distance as compared to the capabilities of other cartridges. In testing in Idaho, an observer at the target could not see the shooter in the open on the desert floor at eighteen hundred meters. Adding a suppressor, the sniper—with no camouflage—could not be seen at this distance even with binoculars.

The rifle repeats its accuracy and holds its zero very well. The system can be disassembled and reassembled with no change in the zero. This includes removing the barrel, optics, and suppressor, then putting everything back together and shooting again. We have seen this capability with the CheyTac and the Accuracy International systems.

The CheyTac bullet has excellent penetration properties. It can penetrate Level IIIA armor at 1,800 meters and a cinder-block wall at 450-plus meters. It will penetrate 2.5-centimeter cold-rolled steel at 180 meters and 1.25-centimeter cold-rolled steel at 750 meters. The .408 system as an antipersonnel weapon is limited only by flight time to its intended target. In testing at the Yuma Testing Grounds, potentially lethal engagements of

CheyTac® M-200

The Reference Standard Bolt Action Rifle in the 408 CheyTac® Caliber

Features:
- •CNC Machined, Receiver
- • Attachable Picatinny Rail M-1913
- • Detachable Barrel
- • Integral Bipod
- • 3.5 lb. Trigger Pull
- • Highly Effective Muzzle Brake

Specifications:

Action Type	• Ultra Heavy Bolt
Caliber	• 408 CheyTac®
Sights	• None, Scope Rail Provided
Overall Length	• 55 inches (stock extended)
Barrel Length	• **29** inches
Magazine Cap.	• 7 Rounds
Weight	• 27 lbs.
Stock	• Retractable

≈ CheyTac M-200 (*CheyTac*)

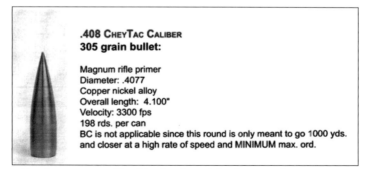

.408 CHEYTAC CALIBER
305 grain bullet:

Magnum rifle primer
Diameter: .4077
Copper nickel alloy
Overall length: 4.100"
Velocity: 3300 fps
198 rds. per can
BC is not applicable since this round is only meant to go 1000 yds.
and closer at a high rate of speed and MINIMUM max. ord.

≈ For closer ranges, CheyTac developed the faster and
flatter-shooting 305-grain. (*CheyTac*)

**.408 CheyTac Caliber
419 grain bullet:**

Magnum rifle primer
Bullet diameter: .4077"
Copper nickel alloy
Overall cartridge length: 4.307
Velocity: 2900 fps
198 rounds per ammo can
Ballistic Coefficient: .94 (avg. over 3500 yards)

⋩ **The CheyTac .408
standard bullet (CheyTac)**

simulated soft targets (gelatin) were consistently made even at subsonic velocities because of the projectile's excellent stability through the transition into subsonic velocity.

As an antimaterial weapon, it's also very effective. The .50 BMG has an initial higher muzzle energy of about 11,200 to 12,000 foot-pounds, as compared to the .408's muzzle energy of 7,700 foot-pounds, However, at about 630 meters, the remaining energy of the .408 is higher than the .50. In fact, CheyTac claims that the 419-grain .408 projectile will defeat any material that the .50 BMG can, except for those targets that require an explosive projectile, such as the Raufoss round (described below). Materials such as jet engines, engine blocks, and surface missiles can easily be engaged and defeated with solid projectiles such as the CheyTac copper-nickel-alloy bullet.

The .50-Caliber BMG

The .50 BMG was originally developed in World War I as an anti-material round. Shortly after the Korean War, people began experimenting with developing a shoulder-fired rifle that could safely handle the awesome power of this round, which was initially meant for weapons mounted on a vehicle or used with a heavy tripod. Early results were not very good, but over time the .50-caliber BMG ammunition and rifle design evolved to sniper-level accuracy. Civilian competition shooters with hand-loaded ammo have shot .25- to .5-MOA groups at nine hundred meters—meaning groupings of one to two centimeters—not bad at that distance!

Like the CheyTac bullet, the rounds in these world-record shots are solid bullets that are turned on a lathe. The downside of these solid projectiles is faster barrel wear and, of course, cost.

Standard military .50 BMG ammo is not that accurate. This is not surprising, given the fact that the bulk of military .50-caliber ammo is manufactured to be used in a machine gun, which is designed to shoot a pattern, not a sub-MOA group. Newer, more accurate military ammunition is available and, indeed, there have been several extreme long-distance confirmed kills made with .50-caliber rifles in the Iraq and Afghanistan theaters.

One of the newer, more accurate rounds is known as the saboted light armor penetrator (SLAP) round. A sabot (or shoe) is a sleeve, usually plastic in rifles, that surrounds a bullet that is smaller than the actual diameter of the barrel, enabling the round to exit the barrel at much higher than normal velocities. After the round leaves the barrel, the sabot falls away and the smaller, lighter round rockets toward its target.

≈ **Example of a sabot plastic sleeve**

The reported muzzle velocity is close to 4,000 fps and has the capability of penetrating two-centimeter steel at 1,350 meters. Shooting with that kind of muzzle velocity produces a nice, flat trajectory with a low max ord, and that sort of penetrating power at distance makes it extremely dangerous.

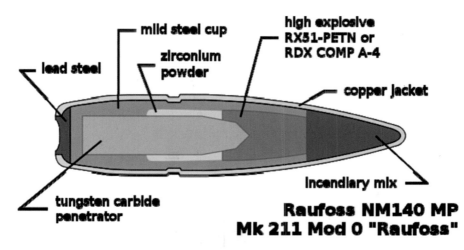

mild steel cup

zirconium powder

high explosive RX51-PETN or RDX COMP A-4

lead steel

copper jacket

tungsten carbide penetrator

incendiary mix

Raufoss NM140 MP Mk 211 Mod 0 "Raufoss"

≈ **(Michaelis, The Complete .50-Caliber Sniper Course)**

Another .50-caliber round is the Raufoss Multipurpose round; this has been called the crown jewel of .50 caliber.

Developed in Norway, the Raufoss incorporates an exploding, armor-piercing tungsten carbide penetrator. During the bullet's acceleration to the target the incendiary compound compresses, which makes a small air pocket. Upon impact the air pocket is compressed, leading to the ignition of the explosive mixture, which sets off a tiny explosive charge. As the tungsten penetrator jolts and fragments out of the bullet's core, white-hot sparks of

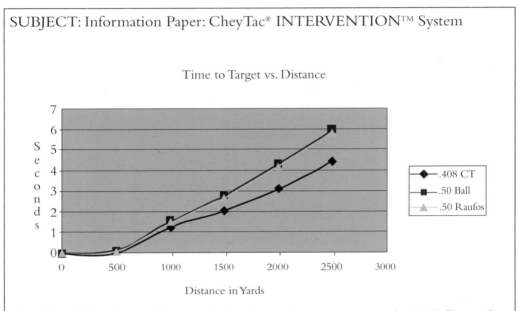

SUBJECT: Information Paper: CheyTac® INTERVENTION™ System

Time to Target vs. Distance

Note: The .50 Raufoss cartridge was designed as an improvement over the .50 Ball cartridge. Data indicates that kinetic energy profiles are very similar.

At 1000 yards the .408 CheyTac® projectile exceeds the .50 BMG projectile by approximately 1 second and at 2500 yards (1.42 miles) by approximately 1.5 seconds.

⌃ CheyTac information paper (CheyTac)

zirconium particles follow, capable of igniting fuel or explosive vapors in the area of bullet contact.

According to its developer, the end result is the Nordic Ammunition Company (NAMMO) equivalent firing power of a 20mm projectile, and due to its penetration and delayed detonation it moves projectile fragmentation and damage effect inside the target for maximum antipersonnel and fire-start effect. The Raufoss bullet also happens to be very accurate.

In Dean Michaelis's book *The Complete .50-Caliber Sniper Course*, he notes that even if you are a great shot with a lower-caliber sniper rifle, technique becomes paramount as you step up to the big dogs. The .50-caliber rifles must be shot neutral, meaning that you have to have your natural point of aim, be straight behind the gun, and not try to muscle the gun to your intended target. Trying to force the gun will result in a miss. Finesse and technique will bring you home with a hit.

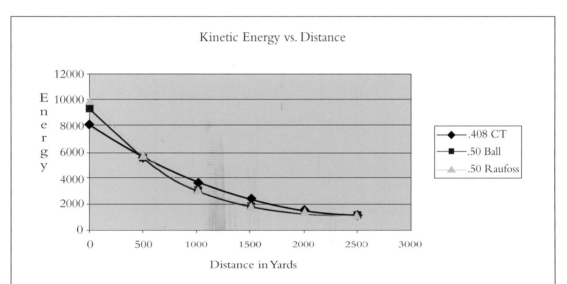

Note: The .50 Raufoss cartridge was designed as an improvement over the .50 Ball cartridge. Data indicates that kinetic energy profiles are very similar.

The .408 CheyTac® kinetic energy surpasses the .50 Ball and the .50 Raufoss at approximately 400 yards and maintains the lead beyond 2500 yards (1.42 miles).

In supervised tests, the .408 CheyTac® projectile penetrated armor and laminated glass that was resitant to the .50 BMG projectiles (*US Armed Forces Journal*, August 2003)

⌃ *(CheyTac)*

The U.S. military has the following .50-caliber cartridges in its inventory:

- M33 ball (not very accurate)
- M17 tracer
- M8 armor-piercing incendiary (reasonably accurate, reportedly)
- M20 armor-piercing incendiary (a tracer shot for the M8)
- M903 saboted light armor penetrator (SLAP) (composed of a 350-grain tungsten steel penetrator, this round requires a different twist rate than other standard rounds)
- M933 SLAP tracer (a tracer for the M903)
- Mark 211, Mod O (developed for and adapted by the U.S. Navy Special Warfare snipers, this round is also referred to as a greentip)

The selection of different cartridges depends upon mission requirements; each round has a different ballistic profile, which the shooter needs to know. For longer distances (nine-hundred-plus meters), where the .50 caliber can be employed, proper analysis and accurate adjustment for meteorological and environmental fac-

.50-Cal AP Round Short-Range Media Penetration (in inches)

Medium	200 meters	600 meters	1,500 meters
Sand (100 lb dry weight)	14	12	6
Clay (100 lb dry weight)	28	26	21
Concrete	2	1	1
Armor plate (homogeneous)	1.0	0.7	0.3
Armor plate (face hardened)	0.9	0.5	0.2

tors become critical in ensuring accurate hits. The use of a ballistic chart or software program and the input of environmentals and accurate ballistic data can ensure a first-round hit.

CheyTac® / 408

HIgher Performance Is The Future of Ammunition

CheyTac® 408/419gr. (M40) vs. Current Issue Military M8 AP
Range 650 Measured Yards
Steel Plate 1/2 Inch Thickness

While the 50 cal. M8 AP ammunition did create an opening in the plate at 650 yds., it is clear that the round did defeat the plate. The penetration is dramatically sub caliber and the front of the plate shows the projectile's energy dissipation. The 408, 419 gr., M40 round shows a full caliber. clean penetration at the same measured distance. The interesting point is the M40, 419 gr. 408 CheyTac® round is not primarily intended as an AP round. It is the standard, long range, soft target interdiction round. CheyTac® produces multiple types of AP rounds for the 408 which have added and unique capabilities for the Military Professional.

The M40 .408, 419 gr. will defeat 90% of the targets that the average sniper is likely to see, whether it be steel, glass, ceramic, or other materials. The only things that the M40 won't defeat are Rolled Homogeneous Armor and 4" armored glass. It should be noted, even 50 cal AP ammo will not defeat these targets. If the operator has to face this kind of threat he will need a true anti-tank munition.

The only types of targets which could present this armor profile is a heavy armored car or truck. 408 or 50 BMG will not be sufficient to stop these targets cold. A LAW or other anti-tank weapon would be needed.

The critical issue in ammunition and weapons facing the dedicated sniper operator today, is performance. At CheyTac®, we are dedicated to the highest level of performance in our design and manufacturing. This exacting level of performance is your advantage in the real world of opeations.

⌃ **Some serious penetration from CheyTac** (*CheyTac*)

Look at the penetration from the chart on page 154 and realize that these are only specs for the standard AP round, not the SLAP or Raufoss. Now you can understand why the .50 may not be the best choice to use when concerned about overpenetration. You don't want any friendlies in the house three blocks away from the target to get hit as well as your bad guy.

.50-Cal AP Round Media Penetration at 100 yards (in inches)

Medium	Inches
Concrete (solid)	9
Timber logs	96
Steel (non-armored)	1.8
Aluminum	3.5
Tamped snow	77
Dry soil	28
Wet soil	42
Dry sand	24
Wet sand	36
Dry clay	42
Wet clay	64

⚐ (*Data from Plaster,* **The Ultimate Sniper**)

For the eighteen-hundred-plus-meter shot in the rocky mountains of Afghanistan, though, this dog will definitely hunt.

Sniper-Grade Bolt-Action vs. Semiautomatic Rifles

Here, in a nutshell, are the pros and cons of the sniper-grade bolt-action and semiauto weapons:

Bolt-Action: Advantages

- *Degree of accuracy.* Typically .5 MOA.
- *Very reliable.* Rare malfunction, secondary to design and number of parts in the firing mechanism. The only real parts of the rifle that are going to fail in the short term are the trigger and firing pin; if using low-grade or bunk ammunition, the bolt could lock up.
- *Ease of field maintenance.*

Bolt-Action: Disadvantages

- *Rate of fire.* Depending on the manufacturer, the rifle may have an internal magazine, which in the lighter-class rifles limits the shooter to five rounds. Some manufacturers, such as CheyTac, Accuracy International, and McMillan, have addressed this issue by designing their rifles

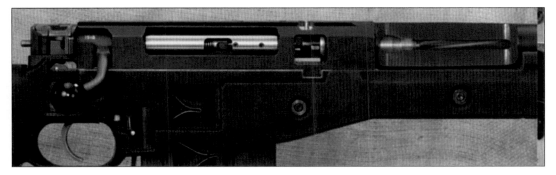

⌃ Cutaway of an Accuracy International rifle receiver and bolt (*Curtis Prejean, courtesy of Accuracy International*)

to use detachable box magazines that can hold more rounds and be exchanged for a fresh magazine as needed. The CheyTac magazine will hold seven rounds; typical .308 magazines will hold up to ten. Despite this increased ammunition capacity, a bolt-action rifle will never have the same rate of fire as a semiautomatic because working the bolt manually takes some time.

Semiautomatic: Advantages

- *Rate of fire*. The semiautomatic action will cycle the bolt much faster than we humans can ever hope to.

Semiautomatic: Disadvantages

- *Degree of accuracy*. The best semiautomatics will shoot 1 MOA in the lighter calibers. In the heavy-caliber rifles, such as the Accuracy International .50 caliber, the accuracy is closer to 1.5 MOA. Granted, this is very good, but, depending on the mission and the target, it may not be good enough.
- *Dependability*. There are many more moving parts in the semiautomatic rifles. Any of these parts can fail. Semiautomatics also have a higher chance of a feeding/ejection malfunction than a bolt-action gun.
- *Ease of maintenance*. More moving parts means more to clean; in gas impingement semiautomatics, carbon buildup is another thing you need to keep a close eye on.

What Can Mess You Up: Accuracy Issues

Anyone passionate about shooting wants to do it well, and doing it well means accuracy—and maybe throw speed in there, too. So many factors can influence shot placement, but they may not all be in the forefront of your mind.

Here are some considerations beyond environmentals and ballistics to think about:

Barrel wear. Hopefully, you have been keeping a record of the number of rounds you have put through your rifle. (For a sample Barrel Log, see Appendix A.)

> Brandon's .300 Win Mag with NF optics; with a 2,900 fps muzzle velocity, this is a real tack-driver out to nine hundred meters.

Over time, barrels lose their accuracy. This is secondary to the degradation of the rifling and the throat of the barrel. The throat is the part of the barrel that takes the brunt of the high pressure and heat of the round being fired.

Some rifle cartridges wear faster than others. In general, hotter loads have a shorter barrel life than the slower rounds.

A loose scope. The repeated stress of the rifle's firing can loosen the scope mount just enough that the scope can move slightly on subsequent shots. Unless you catch it ahead of time, that will ruin your day.

A damaged crown. Again, the crown is the last metal part of the barrel that the bullet touches before it escapes the rifle.

As the bullet leaves the barrel, all the hot gases that have been pushing it down the barrel also escape, and if they do so evenly around the bullet's base (because of damage to the crown), the resulting uneven pressures can influence the bullet's flight and stability, affecting its accuracy.

If the crown gets chipped or worn, your rifle's accuracy can suffer. Usually you see flyers in your groups, that is, individual shots that land way outside your grouping.

Parallax problems. To see parallax, set your rifle up on sandbags so that it is solidly placed. Run your scope up to max power and line up the crosshair perfectly on a bull's-eye at ninety meters. Without touching the rifle, look at the crosshair and move your eye and head very slightly, both side to side and up and down. If the crosshair moves from the bull's-eye, you have a parallax problem.

This can be corrected by the scope manufacturer, but it must be corrected for a specific distance. Most sniper-quality scopes have a parallax adjustment knob.

Fouling of the barrel. Copper fouling is a close second to loose scope-mount screws.

Copper fouling occurs slowly and is hard to see and pinpoint. What you will typically notice is that your groups start opening up over time.

If you determine that copper fouling is a problem, it is fixable with available copper-removal chemical solutions. Be careful with the aggressive solutions. If left in contact with your barrel for too long (fifteen minutes or more), they can damage the steel as well as remove the copper.

Also, be careful not to let these chemicals get into the trigger assembly of your rifle.

Problems with your vision. Stay on top of it, boys and girls! Wear your eye protection religiously, and don't be too proud to stop in and say hello to your local optometrist now and again.

8

SUPPRESSORS

≽ SEAL sniper on overwatch with a suppressed MK-12

During Glen's third phase of Basic Underwater Demolition/SEAL (BUD/S) training, he made several extra laps in the freezing cold ocean off San Clemente Island as punishment because he couldn't stop himself from calling a magazine a "clip."

Too much *T. J. Hooker* as a kid, I guess.

In the same way, most of the general public not in the know refer to a suppressor as a "silencer," the idea being (I guess) that a long-barreled rifle fitted with a "silencer" would make no more noise than your one-year-old passing gas.

Which of course is not even vaguely true.

There really is no such thing as a *silencer*, so from here on out, it's *suppressor* only, understand? Now, this is just one guy's opinion and is not shared by the U.S. Department of Justice, or by the ATF, for that matter, who still refer to the mechanism as a *silencer*. We're trying to start a grassroots movement. Roger? Okay.

So what exactly does a suppressor suppress? Pretty much everything coming out the end of the barrel. Gas, flash, noise, speed . . . *everything*.

The modern suppressor dates back to the early twentieth century; its development is credited to an enterprising American inventor named Hiram Percy Maxim, who also helped develop mufflers for cars (not to be confused with his father, Hiram Stevens Maxim, who invented the first portable, fully automatic machine). From a physics standpoint, the two

An early brochure advertising Hiram Maxim's suppressors ≳

« Traditional "can"-style suppressor; this is a Gemtech HVT model, and it is a great suppressor to use on both semiautos and bolt-action rifles. (*Redstick-Firearms.com*)

inventions work on a lot of the same principles, and back in the day people called suppressors "firearm mufflers."

There are two types of suppressors, commonly referred to as cans and integral suppressors. Cans are attached to the end of the barrel, either threaded on or mounted in a variety of other ways. Integral suppressors are built right around the barrel of the weapon itself—integrated into the guns design. Both employ the same principles to get the job done.

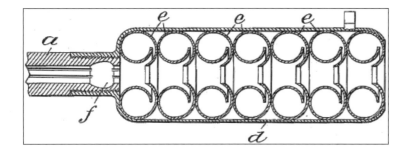

« Hiram Maxim's patent drawing for his original suppressor (*U.S. Patent Office*)

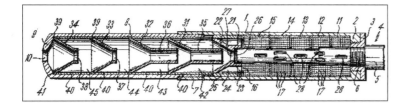

« Diagram from a 1971 Heckler & Koch patent; look to the right of the diagram and notice that initially the barrel is vented out to a wire mesh pack.

There are two factors that cause a gun to make a lot of noise. First is the release of gas pressure from the barrel. When the powder from the cartridge explodes, the bullet gains speed as it heads for the emergency exit. As the bullet clears the end of the barrel, the pressure from the gas behind the round needs to go somewhere, and that explosively rapid expansion is the first sound you hear as the shooter (not counting the explosion of the primer when the firing pin strikes it).

Shooters Depot Advanced Rifle Integral External Suppressor (ARIES) has been called the "holy grail of silencers." It uses carbon fiber tubing. (*Shooters Depot*) ⩶

The second sound is the ballistic crack of a supersonic round as it breaks the sound barrier (approximately 1,100 fps, depending on temperature) leaving the barrel.

Actually, when you are the trigger man, the muzzle blast and ballistic crack come so close together that they essentially become one. If you're downrange from the shot, though, depending on the weapon and ammunition, you'll most likely hear the ballistic crack *before* the report of the weapon at the muzzle.

Advanced Armament Corp billboard; AAC makes fantastic suppressors that can be found on a lot of weapons. Suppressors are legal in thirty-six states. (*ACC Remington*) ⩶

That is, you will if you're lucky. If you don't hear the crack while being shot at, most likely you have a hole in you somewhere that needs tending to. (When using subsonic ammunition, there is no ballistic crack to hear.)

A suppressor works on a couple of principles. It's all about gas pressure management.

First, holes or baffles in the barrel or in the can create more empty space the exploding gas can expand into before reaching the end of the barrel. The baffles are separated by spacers and employ different patterns and designs to direct the gas to the expansion chambers. A suppressor allows the propellant gas to dissipate, providing more

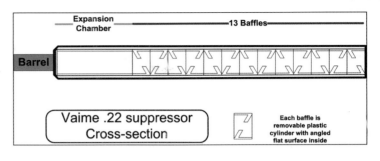

Vaime .22 suppressor Cross-section

Each baffle is removable plastic cylinder with angled flat surface inside

↟ **Note the baffles and expansion chamber.**

volume to be filled. More volume equals less pressure, hence a lower volume at exit.

In addition, the baffles and packing material built into the suppressor allow the gasses to cool, and the lower temperature also lowers the pressure. Depending on the model suppressor, some are packed with fiber-based material, so they can be shot wet. Dunking the suppressor in water prior to shooting will cool the gases down even more, further reducing the sound pressure. Some pistol suppressors are designed to be shot wet and may come prepared with some other liquid in the fiber to aid in the cooling of the gasses, perhaps oil or some type of gel. These will last only a few rounds and are often unreliable and messy.

Other packing material can be used as well—steel wool or some other mesh-type material. These will most likely have a longer life span than the fiber but can't be used wet.

Finally, the baffles also provide the propellant gasses space to roam around in, creating traps and turbulence so that some of the gas is delayed and reaches the end of the line more slowly than it would without the suppressor.

By increasing the time the gas has to reach the end of the suppressor, you are changing the energy released, and the noise created by the pressure will be different, more like dropping

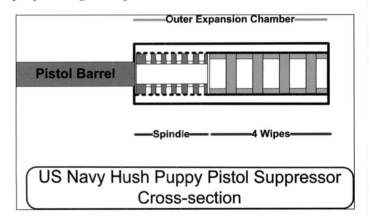

US Navy Hush Puppy Pistol Suppressor Cross-section

The Hush Puppy was a classic Vietnam-era Smith & Wesson Mk22 Mod 0.9mm equipped with a silencer. You can see from the photo that it uses a "wipe" system, which (exactly like baffles) creates multiple chambers for the gas to expand, cool, and delay prior to exiting the barrel. The wipes would either be full sheets, be stamped slightly, or have a hole prepunched for the bullet path; they would need to be replaced frequently, which makes this system relatively obsolete for the modern shooter. ↟

Another of SD Shooters Depot's amazing carbon fiber suppressors. Unlike most suppressors, with SD Shooters Depot's patented Advanced Rifle Integral External Suppressor (ARIES) System, there is no point-of-impact change when using the suppressor. »

a copy of the Yellow Pages on a tabletop than the telltale BANG we all know and love.

Most modern suppressors are not designed to be shot wet anymore and no longer have any features that require regular maintenance or detract from the life of the unit. Twenty-first-century models use unique designs to control gas pressure and

AWC's Ultra II Bolt action integral sound suppressor; AWC uses the latest technology to get match-grade-quality fully suppressed barrels. Super-quiet, and super-impressive. (*AWC*) »

also experiment with controlling the actual frequency of the muzzle blast by using such cool technologies as phase cancellation and frequency shifting.

Suppressors aren't just for the James Bond moments, where you are using a single-action .22 for a whisper-quiet assassination of some clown in a tux. There are many practical reasons for their use in the field.

First and foremost, they protect your hearing, not only in the moment but in the long term as well. If OSHA had any say in the matter, every military weapon would have a suppressor attached.

The ability to hear when operating in any situation, the ability to maintain command and control and keep one of your most crucial senses keen for situational awareness in a combat situation cannot be stressed enough. Shooting even a 5.56mm round without hearing protection will leave your ears ringing; get up to the larger calibers and you're doing permanent damage. Take

Glen at sunrise in Tikrit with his suppressed SR-25 ⌄

weapons indoors and the noise just seems to keep getting louder, reverberating off the walls.

Suppressors were used effectively in Vietnam when taking out sentries, dogs, and other critters used to sound the alarm, and are prevalent in the combat theaters of today.

One suppressed pistol specifically designed for the SEALs in the 1970s was called the Hush Puppy. I can attest that there are plenty of mongrel dogs in Iraq and Afghanistan that fell to suppressed weapons in order to ensure mission success.

(Before you jump all over me, by the way, I'm a dog lover, too, and as I document in my book *The Red Circle*, our platoon took in some abandoned dogs in Afghanistan and even brought them home to the States. But in war, casualties are casualties.)

At night, using a suppressed sniper rifle also makes your flash signature all but disappear. Until you see it with your own eyes (or *don't* see it, I guess it's more accurate to say), it's truly hard to believe how invisible you become.

We were training in northern Washington State in 2002, doing an advanced sniper course at a ranch called Bull Hill, and we did an exercise where half the crew was downrange in the prone position, watching the tree line some 360 meters back. In the trees were shooters using the SR-25 with attached Advanced Armament Corp suppressors, putting rounds on target beyond and above the observers.

As individual rounds were fired, we did our best to locate the shooter. It was all but impossible to locate the muzzle flash. (Although it was good training to hear that telltale *crack* over our heads, and it also gives you a warm and fuzzy feeling to hear your buddy pounding a head-sized steel plate from 540 meters out at night.)

When using large-caliber rifles, especially when shooting from the prone in a dry or dusty area, a suppressor significantly lowers the amount of dust that kicks up from the muzzle blast, ensuring that you aren't compromising your hide and allowing you to possibly take a follow-on shot if necessary.

For any unit mounting raids in a crowded neighborhood, hitting a house with suppressed weapons will keep the neighbors wondering what is going on, as opposed to having them instantly aware that there's a gunfight next door.

That just amounts to one less thing to worry about.

With any rifle, using a suppressor also reduces recoil substantially, which allows for a quicker follow-through and acquisition of targets. The importance of this benefit cannot be understated, especially with high-power rifles like the .338 Lapua and the .50 calibers.

If and when you are lucky enough to be able to engage multiple targets in quick succession, lessening the dust cloud kicked up, the sound signature, the muzzle flash (especially at night), and the recoil are all going to lead to a much faster second shot. The dramatic change in the sound emitted after your weapon is fired will also leave the enemy confused and unable to pinpoint your location.

As the Finnish say, "A silencer does not make a rifleman silent, but it does make him invisible."

It was the spring of 1999, and we were out at what we still refer to as "Shaw's," after famous shooter John Shaw, formally known as the Mid-South Institute of Self-Defense Shooting.

We were working in the kill houses doing close quarters battle (CQB), also termed *close quarters combat* (CQC) or more simply *room clearing*—"a flurry of asskicking," as our chief liked to call it—working the flow of a squad or platoon through a house with live ammo. One of the takedowns we did was completely suppressed.

At the time our sidearms were HK Mk 23s, basically the military version of the Heckler & Koch USP .45 with their barrels threaded for suppressors. I loved these pistols. They were total hand cannons, and very accurate. Some of the boys didn't like them because they were big; if you didn't have large hands, this weapon could feel unwieldy. With suppressor attached, the thing felt so big you practically wanted to attach a buttstock to it and use it like an SMG.

On these surprise hits, instead of initiating the assault by breaching the door with demo, the whole operation is quiet. There's no loud yelling, "Clear right, clear left, all clear." Targets are all taken down by suppressed pistol—at least until your cover is blown, at which point you go overt. It's the report of that suppressed pistol I remember so well. Even after using it on the range, in the house it just echoed differently. It sounded a lot like someone holding a large phone book about three feet off a table and dropping it flat onto the floor.

It was nowhere close to silent, but when I put myself in the enemy's place, perhaps sitting in a back room and not knowing what was coming, the noise sounded random and unidentifiable. I might have perked up, knowing that I'd heard *something*, but would definitely not have associated that *something* with the crack of a pistol—at least not at first.

⨠ A SCAR light suppressed. At 3.5kg without suppressor or glass, this rifle is easy to carry.

9

OPTICS, BALLISTIC SOFTWARE, AND OTHER COOL STUFF

Advancement in optics and the integration of technology inside the optics is the way of the future. We have scopes under development right now that have computer and software integrated directly into the scope.

The sniper scopes of the twenty-first century are capable of seeing through smoke, fog, and rain and can identify targets through buildings at night. They can also adjust for wind and calculate complex ballistic firing solutions that take into account muzzle velocity, environmental factors such as temperature, altitude, humidity, barometric pressure, and even how the spin of the Earth (Coriolis effect) will alter the bullet's flight path.

Rifle Scopes

There is an abundance of great companies producing cutting-edge optics for weapons of the twenty-first century: Leupold, U.S. Optics, Nightforce, Zeiss, Swarovski, Schmidt & Bender, Bushnell, Burris, and Nikon, to name just some of the most prominent. When it comes to choosing optics for your sniper rifle, any one of these companies can provide you with a scope to suit your needs and successfully complete your mission.

If you're building your own weapon, consider spending as much as you can afford on good optics, scope rings, and mounts. It will make not only a critical difference in shots on target at long ranges, but it will also reduce eye fatigue, improve clarity, and potentially enhance reliability.

To demystify scope specs a little: A "2.5–16x 42mm" scope is one with adjustable magnification, from 2.5x to 16x, and a front (objective) lens diameter of 42mm. To see why any of that should matter, let's get inside the scope for a moment and break it down.

There's a fair amount of debate on the blogosphere over how relevant the phenomenon known as Coriolis effect is to a shooter's accuracy. In layman's terms, the Coriolis effect refers to the impact of the Earth's spin on a bullet's trajectory in long-distance shots.

Here's the bottom line: Shooting inside 720 meters, you don't need to take this into consideration at all. When you're making shots out beyond 1,800 meters on targets as small as a half-starved Afghani fighter, though, everything counts.

At the poles the Earth's rotational speed is minimal, but get to the middle latitudes and our little ball is spinning at about thirteen hundred kilometers per hour. (Head to the equator and that's more like sixteen hundred kilometers per hour.) That's moving.

What people typically forget to take into account about the Coriolis effect is that how much the Earth's rotation is going to affect your point of impact depends on which direction you're shooting in—that is, what compass bearing. If you're shooting directly north or south, it will have a much larger impact than if you are shooting due east or west.

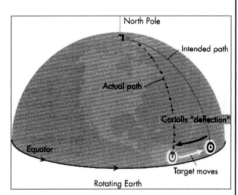

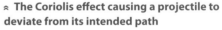

≈ **The Coriolis effect causing a projectile to deviate from its intended path**

Let's say you are shooting a .50 caliber at a compass bearing of 180 at 2,250 meters. With no wind, the untrained shooter will hold for a point-of-aim point of impact (all other environmental factors taken out of the equation). The bullet has a muzzle velocity of approximately 2,800 fps; at mid-latitudes, the Earth is spinning at about 1,200 fps.

That means the Earth is definitely going to move a little during that 2.5-second flight time—with your target on it! If you don't account for Coriolis effect, when shooting due south your bullet will miss to the right; if shooting due north, you'll miss to the left. Savvy?

Fortunately for me, I'm Irish, so I don't have to worry about any of this. I just get lucky.

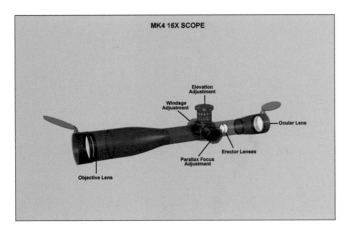

MK4 16X SCOPE

Elevation Adjustment

Windage Adjustment

Ocular Lens

Erector Lenses

Parallax Focus Adjustment

Objective Lens

« **Leupold MK4 16x scope**

Starting from the dangerous end of the rifle, light entering the scope moves toward your eye and is gathered up by the objective lens. This lens might be inside a cover, recessed like a frightened turtle, and it may also have an anti-reflective coating on it. The cover helps to prevent glare, which lowers the chances of the shooter giving away his position; the coating allows more light to be gathered into the lens and may also contribute some antifogging properties.

The bigger the objective lens, the more light it can let in. More light is good. In addition to making things look bigger, the process of magnification also reduces the amount of light inside the tube, so the higher the power of magnification, the more you need that big objective lens so that your image isn't blacked out. A magnification of 4x will quadruple the size of the image; 9x is nine times bigger—easy, right?

After the image makes its way to the inside, the light waves get bent at the lens; somewhere in the tube the image gets flipped upside down and backward. Because of this, an erector lens

SEAL with a standard M-4 equipped with EOTech red dot optics and M-203 40 mm grenade launcher ⍥

or erector cell is used to correct the image, so that it appears properly to your eye. With variable power scopes this cell will be contained together in a fixed tube, and this tube moves as a unit as you adjust your magnification. The reticle will typically be attached to the back portion of these internal lenses.

Mil-dot reticles can be illuminated for night shooting. ⩔

Finally, your image comes to you through the ocular lens and provides the exit pupil, which is the size of the column of light that leaves the eyepiece on your scope. The size of the exit pupil is determined by dividing the objective lens size by the magnification; for example, a 5x 40 will give you an 8-millimeter exit pupil. The smaller the exit pupil, the closer your eye has to be to the ocular lens for a proper view. Too small and you'll be right up on it, getting smacked in the face every time you pull the trigger.

The size of the exit pupil is also important in that it gives you more wiggle room when finding the appropriate eye relief, which is the point at which all the scope shadow disappears and you are looking at a full tube's view of a bright, crystal clear image.

Glen keeps warm with some aftermarket ghillie product. Don't forget to cammie and veg up your scope and weapon. ⩔

Adjusting windage (compensating for the effects of wind on your bullet's trajectory) and for elevation moves the internal lens assembly in relation to the outer lenses, so that your point of impact will adjust ever so slightly with those clicks.

This is why the nicer optics will provide .25-MOA adjustments. Lenses that efficiently transmit the light to your eye will provide you with the brightness that you want, and high-quality, properly ground lens glass will give you the clarity you need to properly ID your target.

However, even with one of the most expensive scopes, if you don't how to mount it to your weapon properly and set it up for your eye relief and body size, then you may as well stick with iron sights.

While teaching long-range shooting courses, Glen and I have both seen well-heeled civilian shooters show up with the latest Accuracy International .308 and a top-of-the-line Leupold Mk IV scope, complaining that they just can't seem to get any consistency out of their groups. A quick shake of the scope itself will typically find it loose, with the mounting bolts not even tightened down. As the old adage goes: "If you're going to be stupid, you'd better be tough."

Two other common setup errors we saw were failing to set natural eye relief and failing to set the reticle in a perfect plane relative to the weapon. (If it weren't for the fact that most men hate reading instructions and inherently know how to do every-thing, we wouldn't see so many of these problems on the range.)

Before doing anything else, you have to find the appropriate eye relief. Start with the scope toward the dangerous end of the weapon and get in a comfortable prone shooting position with your eyes closed. Once you're settled, open your eyes and move the scope into a position where scope shadow is eliminated. Try other shooting positions to ensure that the location is correct, then lock those bolts down.

Essentially, for a traditional mil-dot scope, when on target the reticle should look like a T, not an X. Simply loosening the scope rings and rotating the scope will quickly fix that. It's hard to have an accurate wind hold and stay on target when your reticle is in the X position; your vertical point of impact will take a beating.

Next up is focusing the reticle. In order to get the reticle crisp, point the scope at a white background, then rotate the eyepiece until it is in perfect focus for your eye. If your scope has an eyepiece lock ring, after getting the reticle focus dialed, lock that baby down, too. Understand that as you get older and your eyesight deterio-rates, somewhere down the road you may need to readjust this.

One of the most popular scoped optics employed by the Spe-cial Operations community is the variable-power Nightforce with

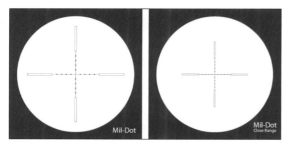

> Two Nightforce reticles using the classic mil-dot system military shooters are trained on (*Nightforce*)

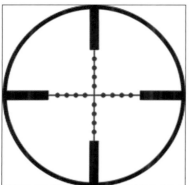

> Mil-dot reticles can be illuminated for night shooting. (*Nightforce*)

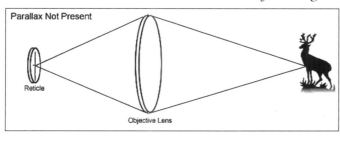

> (*U.S. Optics*)

the illuminated reticle. The variable power allows for a lot of flexibility between rural and urban operations, and the illuminated reticle enables the sniper to operate at night in an urban setting where ambient light is available without night vision.

Many an Al Qaeda operative has fallen prey to the Nightforce scope and that illuminated reticle. While superior optics exist, you cannot beat the Nightforce for price and reliability.

Parallax

The majority of scopes used for long-range shooting have three knobs on them in a cluster in the middle of the unit. Two have obvious functions: They adjust windage and elevation. But what about that third knob with the infinity symbol on the left side of your scope? This is your parallax adjustment.

Parallax is the apparent displacement or difference in apparent direction of an object as seen from two different points not on a straight line with the object.

Okay—what the hell does *that* mean? It means your parallax knob adjusts the internal focal planes of your scope so that the reticle and object image are on the same optical planes. If you have parallax, the little picture of your target is being transmitted either in front of or behind the reticle. Adjusting the knob will move that image plane so that it falls exactly where it's supposed to.

Imagine you are a passenger riding in a car; you glance over at the speedometer and it appears to you to read 125 kilometers per hour. However, what the driver sees in his dead-on view is 160 kilometers per hour (which is the car's actual speed). The way your eye lines up with the needle, which is analogous to the scope's reticle, and the numbers on the instrument panel (the target image) line up differently when seen from the different angle resulting from where you are sitting, off to the side. This is not a precisely accurate analogy, but you get the idea.

Here's another way of seeing it: Hold your index finger up at arm's length in front of you, close your left eye, and sight in on something on the wall. Now, without moving your finger, close your right eye and open your left. Your finger (reticle) has jumped off target.

To put it simply, when your target image and your reticle are not in the same plane, you have parallax. With scopes, especially high-powered scopes, this can be a serious problem, which that third knob is designed to remedy. People often describe this dial as an "object focus" or "reticle focus" knob. Not so much, although the parallax adjustment sometimes does make the object go in and out of focus, especially at long ranges.

The parallax adjustment knob typically has distances stamped on it, but they are rarely dead-on. The way to check for parallax is to get in a nice stable position, then move your head slightly up and down or side to side. If the reticle or target appears to move, then you have parallax and need to adjust until that movement disappears.

Remember this, though: *Parallax increases with magnification.* If you are shooting at ninety meters and forgot to adjust for parallax at that range, you are maybe .1mm off inside your scope and have

The top image shows no parallax. In the lower two, you can see that the reticle plane is either in front of or behind the object plane. (Nightforce)

Leica spotting scope fitted with a digital camera; this was a favorite among SEAL sniper instructors.

to quickly send a round downrange at nine hundred meters so you can power your magnification up 15x on your variable power scope. Your image plane will now be 1.5mm off the reticle plane, assuming that you were shooting initially at 1x magnification.

Too nerdy for you? So long as you've come away with an understanding that the knob isn't for adjusting focus, we've done our job here. Good luck.

Spotting Scopes

Modern snipers are going to variable-power scopes because they are much more practical at a variety of ranges. You can dial power back in an urban setting or up in a long-range rural environment. Having a high-power setting also lets the sniper serve as his own spotter, using the increased magnification to get positive target identification or a read on environmental conditions at the target.

Most long-range tactical rifle scopes offer you variable magnification, typically 3.5x to 10x and on the extreme end as much as 8.5x to 25x. The variable power enables the scope to be more practical in a variety of environments. However, that extremely high-powered magnification is not commonly seen on the weapons of choice used by today's operators in most environments.

The ability to really get out there and get a good look at a target is so important that most snipers, especially when operating in two-man teams, bring a high-powered spotting scope with them.

Spotting scopes typically provide magnification starting at about 20x and going all the way to 100x on some of the more powerful models. Most commonly you'll find a 20–60x model slid into the sniper's drag bag.

Let's not forget that one of the sniper's primary jobs is to serve as a trained observer. His ability to stay hidden, observe a target, and relay pertinent information back to command can make the difference in any engagement.

Modern spotting scopes can be hooked up to digital cameras and video recorders, for when that 100x magnification is used for, say, positive identification of high-value targets (HVTs) or to prepare a target package for a follow-on direct action assault of a hard target. Still photos or digital video can then be saved, compressed, and sent via satellite anywhere in the world. I'm sure the president and the joint chiefs have had coffee while watching near real-time video being passed from snipers observing a target in the field on the other side of the world.

(Not to mention that it can also be fun to review your successful long shots again and again and show off to your friends. If you haven't already done so, go to YouTube and check out the video taken through a spotting scope of Canadian snipers taking out Taliban at over eighteen hundred meters with a .50 caliber. Awesome.)

≈ **A small element of SEALs on dusk patrol. Note the sniper (right) with the suppressed rifle. Night-vision optics and suppressors give the element overwhelming advantage.**

The additional magnification is also useful in training on the range and calling accurate wind for your shooting partner. Especially when the weather and wind are up, seeing wind and mirage (the heat-rippling effect you see when you look down a highway on a hot day) is easier with the nicer glass. Then, of course, there's watching what is called "trace."

Simple reticle on a 4x scope; sometimes simple is all you need. ≈

There are longtime shooters who claim to never have actually seen trace, but get on a nice spotting scope behind a shooter on a warm day and you'll get to have your Matrix moment: You can actually see the bullet's pressure wave as it tracks through the air. In fact, you can follow it all the way to the target.

This is where the shooter/spotter pair combination shines. The shooter will call his shot, letting the spotter know exactly what his sight picture was when the surprise break happened, and having actually seen where the bullet impacted from the trace or target splash (the bullet's impact, whether on-target

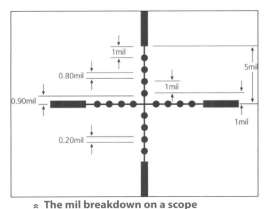

▲ The mil breakdown on a scope

or off-), the spotter can make the necessary corrections for windage and elevation to ensure that any follow-on shots are landing exactly where the pair wants them.

Ranging with Your Scope

Accurately ranging your target is key, and there are many ways to do it. The mil-dot reticle scope is a powerful measurement tool that lets a trained shooter quickly and accurately range for distance and hold for leads by understanding that minutes of angle are visually represented through the dots and spaces between the dots on the scope.

Any good sniper will know this and have his mil-dot leads down cold. The advantage of this reticle versus traditional scope sights is that a sniper can adjust quickly for multiple targets and account for a variety of changing conditions.

In order to calculate a target's range using the mil-dot reticle on

SEAL sniper students shooting an Unknown Distance (UKD) test out to one thousand meters with issued SR-25 sniper rifle and Nightforce optics. »

your scope, keep in mind that 1 mil equals 3.438 minutes of angle, which translates into 3.6 inches at one hundred yards.

For ranging purposes, the only other thing you need to know (or approximate) is the height (or width) of your target. Here is the formula:

$$\text{height of target (yards)}$$
$$\text{x } 1{,}000/\text{height of target (mils)}$$
$$= \text{range in yards}$$

There are tables that can help with quick ranging, as well as useful tools with sliding scales such as the Mildot Master—but it's always good to know how to do it the old-school way, just in case technology lets you down. Batteries don't last forever.

Laser Range Finders

These days, ranging typically involves using a laser range finder. These can be monocular or purchased integrated into binoculars, like Vectronix's Vector,

⌃ **Leica pocket laser range finder; an excellent battlefield-ready range finder**

which we used in the field when I was in the SEAL teams. The Vector has 7x magnification, a built-in digital magnetic compass, a clinometer for all-important angle shooting adjustments, and data links so the information and images can be downloaded if desired. Cool, right? We always thought so.

In the future we will see laser range-finding capabilities integrated directly into your rifle's optics. In fact, it's happening now, though not yet to the level and quality you need when getting out past nine hundred meters.

Modern ballistic software gives the twenty-first-century sniper a huge advantage. A single sniper using a small, handheld computer, ballistic software, and a chronograph (a device that measures the weapon's specific muzzle

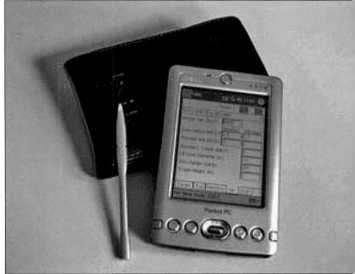

CheyTac developed a very accurate ballistic program after integrating radar data gathered on bullet flights at the U.S. Army's Yuma Proving Ground. One only needs to input a few variables to get firing solutions out past eighteen hundred meters and achieve first-round hits. Their software is shown here in a pocket PC, but it can be used in a variety of handheld devices. (CheyTac) »

velocity—see below), can sight in multiple rifles at the ninety-meter line with only a few shots per rifle.

With the advantage of software and access to the Internet (to obtain environmental conditions and elevation), the modern sniper can be prepared to go anywhere in the world and take his rifle out of the box ready to kill.

A chronograph, which will measure the actual velocity of whatever round you are using, is an important tool in any twenty-first-century sniper's kit.

When using factory ammunition or when reloading, getting a true muzzle velocity for any given round is critical information you need in order to input correct data into your ballistic software to produce an accurate firing solution.

≈ **An Alpha Chrony from Shooting Chrony. (*Tom Gaylord*)**

Just because the manufacturer tells you their ammo delivers 2,780 fps doesn't make it so. It's always best to check.

Consistency is paramount in shooting, and finding and using the same lot or batch of ammunition is important to maintain consistent rounds on target, especially when really reaching out to touch someone.

≈ **Knights Armament makes this Picatinny rail mount (Mil-Std-1913) for the Heads-Up display, their BulletFlight ballistic computer, and another mount for the iPhone. (*Knights Armament*)**

Ballistic Software for the Perfect Kill Shot

When I was attached to the Naval Special Warfare Group One sniper cell in the early 2000s, at one point we were called on to test out some ballistic software applications.

As part of our advanced sniper sustainment training, we would take the guys on several hunting trips each year to keep them on top of their game. Stopping a beating heart in an animal is very similar to doing the same in a human, and these hunting trips were excellent practice.

On this particular trip we were hunting white-tailed deer up in Washington State, right on the western border of Canada, and I had a brand-new .300 Win Mag that was new out of the box. The guys were slow to trust technology, so I decided to do a test run of the ability to sight in a rifle in one environment and then factor in a completely different environment using only ballistic software.

I took the gun to one of the sniper ranges at Camp Pendleton and shot a three-round zero at ninety meters and made the necessary adjustments to my Nightforce scope. At the same time, I shot my zero through a chronograph to measure my .300's muzzle velocity (it was averaging 2,850 fps).

I then went home and looked up the weather conditions for the remote lodge where I would be in a few days and entered that

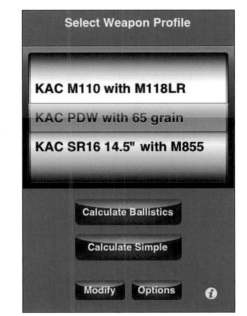

≈ Ballistic software developed specifically for the iPhone; amazing.

≈ Select your weapon and ammunition.

data into my software. I programmed it to give me my come-ups out to nine hundred meters. Three rounds through a brand-new rifle system and I was prepared to kill out to nine hundred meters! You have to love technology.

A few days later, I set out into the mountains in a light rain to still hunt. Deer are like humans and don't like to get wet, and I decided to use this to my advantage, counting on the forecast that the rain would let up in time for the deer to come out and feed before sunset.

« Barrett's Barrett Optical Ranging System (BORS) Ballistic Computer mounts directly on your scope and couples directly to your elevation knob. Set BORS for your specific round and it will automatically compensate for environmentals, distance, and even shooting angle. In the future, expect to see this technology built directly into the scope. (*Barrett*)

After about a one-hour hike, I reached an elevated spot that gave me a great vantage point of a small field, where I had glassed trace with my binos. The center of the field ranged out at about 415 meters. As soon as the rain stopped, there was a mad rush of hungry white-tailed deer, and soon I had a nice-sized buck in my scope. I had already dialed in my elevation for 415 meters and squeezed off a great shot that lifted the deer off its feet and killed it instantly—point of aim, point of impact right through the heart.

To prepare for that perfect shot, I had taken a brand-new rifle, sighted it back in California with three shots, then took

« Input environmentals and get your firing solution; almost takes the fun out of it. Almost.

« The Kestrel 4500 is an anemometer (wind reader) and so much more. It can quickly and accurately give you just about every environmental stat you need: barometric pressure, altitude, humidity, temperature, wind speed, digital compass heading, even wind direction—handy when you need data fast for your ballistic computer.

it to up to a completely different location with radically different environmental conditions including high elevation and a *very* cold climate, where I simply plugged in my new data—temperature, barometric pressure, type of round, and degree of latitude—and with my first shot (including the three test rounds back in California, the fourth shot fired out of that gun since I'd purchased it) I had executed a *perfect* kill shot.

From that moment on I became an evangelist for applying and leveraging technology to the sniper's advantage.

On Ammunition and Ballistics

So many different types of ammunition are in use in the world today. In the history of precision marksmanship, we've evolved from the smoothbore round ball to custom-made tri-metal blends boasting more than 5,000 fps out of the muzzle.

For the layman, and even for many shooters who considers themselves fairly experienced, a discussion of ballistics is often enlightening.

While long-distance shooting is most definitely a complex art, there is no art to ballistics, only science, and a basic grasp of the physics involved throughout the firing and impact cycle makes for a better and more well-rounded shooter. It is essential to have a basic understanding of gyroscopic stability, wind drift, rifling,

Twenty-first-century ammo: LeMas Ltd. is producing some cutting-edge ammunition with incredible results. The authors had the opportunity to demo some of their .300 Win Mag and were blown away by the ballistics, especially at the terminal end: way above and beyond the stopping power of a traditional .300 round. ⌄

The Burris Optics Eliminator Laser Scope (*Burris*) ⌄

↟ **Burris Eliminator; keeps you from having to bring too much shit along with you, which is nice. (*Burris*)**

muzzle velocity, trajectory, Coriolis effect, and terminal ballistics (what happens after the bullet strikes its target).

There is a critical difference between the ammunition that the regular ground soldier is using and the rounds the sniper is equipped with, which is often considered "match grade." What's the difference, and why is it so important? It's all about accuracy and first-round effectiveness.

It's not that these characteristics are not important to every soldier, but they are comparatively more crucial to a sniper—and there are corresponding increases in cost involved. Match-grade ammunition, ammo that is made to a higher standard, costs more and is less readily available.

Attend any long-distance shooting event in the world and most of the small talk you will hear at the firing line is about all sorts of arcane distinctions in custom-made ammunition—different types of powder, custom casings, bullet weights, tricks, myths, and black magic. To those who know, it's the make-or-break difference, the difference between winning and losing a national shooting championship.

For most military snipers, such discussions follow more along the lines of "Who cares?" Custom ammunition isn't part of our world. This isn't to question the authenticity or effectiveness of such hand-loads; it's just that facts are facts. We shoot what we're given—and still achieve grand-champion-like results.

Burris Optics has created a unique system they call the Eliminator Laser Scope. This precision instrument is preprogrammed for more than six hundred commercial cartridges, and unlike all the other systems we've presented in this chapter, the Eliminator is completely self-contained. The scope is a 4–12x 42mm, very respectable in terms of magnification with an objective lens diameter of 42mm.

Its laser range finder won't reach out as far as some of the powerful handhelds; it's only able to reflect off a human-sized target to five hundred meters. So far, anyway. A couple of years down the road, more companies will be following up Burris's lead by integrating everything into one user-friendly scope—range finder, night vision, thermal, laser designators, ballistic computers, and more.

Burris asks you to put in a simple ballistic figure—your bullet drop at 450 meters when zeroed at 90—and then verify at 450 meters. From that point on, all you do is range your target, and the scope will calculate for the angle of the shot and the ballistics profile of your particular cartridge, and put a lighted dot on the vertical post of the reticle where your holdover is. It does not miraculously hold for wind . . . at least not yet.

Night-Vision Optics

"We own the night" is an expression commonly heard when getting tales from overseas. It's so true. From driving to a target you are about to hit totally blacked out, to being able to glass and scan areas completely on night vision, gives the twenty-first-century operator an incredible advantage.

Typically the majority of the platoon or squad is on helmet-mounted NODs (night observation devices) such as the AN/PVS-7, which is common double-eye night vision, or perhaps the PVS-14 Monocular NightVision Device. The latter is better when you're on foot doing long patrols, as it allows you to continue using your natural night vision and depth perception, which helps keep you from falling on your face.

These devices are generally coupled with laser systems mounted on the Picatinny rails of your primary weapon, such as the PEQ-2, which is an IR (infrared) target pointer and aiming light, coupled with an IR illuminator—think Surefire flashlight, only invisible to the naked eye.

SPA-Defense (Simrad) couples their KN203FAB Night weapon sights to these CheyTac sniper rifles. (Bad guys, beware—no more hiding in dark corners.) At 3.5 pounds, though, they add a top-heavy component to the weapon system. As long as you are in a good, supported position, no worries. (*CheyTac*) ⌄

« Nice AR shown here with the PEQ-2 mounted on the top Picatinny rail, with a touch pad switch secured by Velcro on the right side of the forward hand guard. With the Trijicon ACOG sight, this kit will get the job done in a lot of situations.

Insight Technology recently put out the AN/PSQ-23 STORM (small tactical optical rifle mounted), which also has a laser range-finder and built-in digital compass in addition to all the features of the older PEQ-2.

Infrared lasers and visible lasers are good in certain situations (closer range and urban operations), and for those kinds of missions you'll often find a sniper having his weapons system equipped accordingly. But if you are looking to reach out, have a lot of ground to cover, and want surgical precision, there are much better setups available.

The twenty-first century has solved a lot of technological problems that plagued snipers back in the nineties. Today a shooter has a plenty of choices, starting with easily attached models, such as the KN203FAB MK IV or the KN-250, manufactured by different companies but both providing the same type of proven solution. These systems slip into the front of whatever day scope is already mounted, and after a quick boresight on the unit (which is easily done in the field), the weapon system is ready for night operations.

You can also flip the front cover on the unit down and leave it in place for daytime operations, but this tends to be a bit unwieldy. It's better to stow it until the sun starts going down.

These image intensifiers allow you to see your regular reticle and use all the functions of your day scope without putting a whole

« Accuracy International .338 rocking the PVS 26; this setup is fantastic.

new optical unit on there—invaluable when you are relying on consistency and routine prior to taking a difficult shot.

Just as with a pro golfer, there is a program you undertake before every good shot. It's all about the setup. Get your mental and physical plans on the same page, envision a hit, and then go through your pre-shot routine.

If your weapon is fortunate enough to have the full MIL-STD 1913 rail interface (also known as the Picatinny rail), then you have other options available to you.

There are multiple systems that will also mate with the front of your day scope but have their own rail mount system, too, so that they are locked in. Some prime examples are the AN/PVS-22 Universal Night Sight (UNS), the PVS-24, or the PVS-26, which is essentially the same as the 22 but designed specifically for long range. These units attach to the rail in front of your optics and add night-vision capabilities with Generation 3 image intensifiers (which gather and increase ambient light). They can also be installed and removed without any tools.

Another added bonus of the UNS is that it takes no adjustments to get it dialed in to your weapon. It has permanent boresight alignment, which gives you a warm and fuzzy feeling, especially if you have to take a quick shot and it took a lot of maneuvering to get into a good firing position.

AN/PVS-22 UNS seen coupled with a Leupold MK 4 spotting scope with the McCann Industries Optical Mounting System (MOMS) (*McCann Industries*) ≽

Another reason the UNS is a good item to keep in your kit is that it can also be coupled to a spotting scope if you really need to identify targets in a low-light environment. Additionally, it can be used as a handheld monocular device. Sometimes that's really all you need—K.I.S.S. (keep it simple, stupid).

↗ **Optical Systems Technology makes both the MUNS and the PVS-22 UNS sights seen here.**

The newer version of the PVS-22 is called the MUNS (Magnum Universal Night Sight). It operates on all the same principles, but is the latest and greatest (and most expensive) version. It gathers nearly twice the light as the 22 and can engage targets approximately 1.5 times farther than the original UNS. Break out your wallet, Uncle Sam, the boys want the best.

There are thousands of these inline night-vision devices deployed overseas right now, and they are an essential addition to any twenty-first-century sniper's war chest.

Thermal Optics

The twenty-first century just keeps pushing the technological envelope, and the latest thermal imaging/forward-looking infrared (FLIR) sights do not disappoint.

Infrared energy, which we'll call heat, is invisible to the human eye, but these systems enable you to "see" the heat emitted from any object, whether animate or inanimate.

To break out a bit of science: at the lower end of the visible light spectrum is the color red; at the upper end is violet. When you hear people refer to the ever-dreaded ultraviolet (UV) rays, they are talking about the light just above the color violet, which sports a shorter wavelength and harbors more energy. On the other end of the spectrum, just below red, is infrared, and this is where this optics technology lives.

There are two ways IR technology makes it to the markets: uncooled and cryogenically cooled. Uncooled systems are those the operators in the field are using every day. They are less expensive and more durable, but they are also not as awesome as the cryogenically cooled units.

These bad boys' innards are sealed in a system that keeps them at temperatures below 32°F. Operating at these temperatures allows their sensors to produce incredibly clear images, detecting as small as a .2°F change from over 300 meters. Still, they are too fragile to take into the field—for now, at least.

It's a competitive market out there, and several companies are leading the way, making what was once a heavy, bulky, power-sucking device into something rugged, manageable, and able to last in the field. Some of the newer (and incredibly expensive) devices combine image intensifiers, with thermal technology, combining the two images and then overlaying them onto your field of view . . . in real time. Amazing.

The small and very portable CNVD-T (also known as the SU-232/PAS) clip-on thermal device ⩔

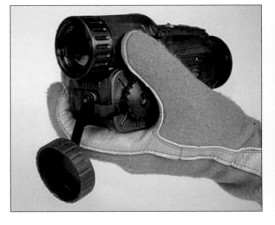

Raytheon and the U.S. Army developed the AN/PAS-13 Thermal Weapon Sight, which along with other companies' FLIR/thermal systems offers some distinct advantages over regular night vision. First and foremost, it operates day or night, picking up even extremely small variations in temperature, and will not be overly affected by rain, smoke, or fog.

Image intensifiers need some light to work, but thermal does not. Imagine being deep into triple-canopy jungle. *Nothing* penetrates that; the night there is like being deep in a cave. In that scenario, hope you brought your thermal goggles.

There are different viewing modes on most thermal devices, enabling the user to view in "white hot" or "black hot" modes. It's like playing around with picture-editing software or reversing a negative—sometimes it's just easier for the eye to recognize a shape or identify a particular target in white or in black. Either way, it's cool.

Also, thermal sights are unaffected when hit by direct high-intensity light, which would cause night vision to shut down. (Actually, night vision doesn't exactly shut down in this case, but *you* do. When you're on a NOD and get hit right in the lens with a powerful light, or if someone or something pops off a flare, you get what's called "bloom," and it will temporarily blind you.)

The PAS-13 comes in a variety of different sizes and strengths: light, medium, and heavy. The light version is for your rifle or carbine and can positively detect targets out just past 450 meters or so. The medium you would see on the squad automatic weapon (SAW) or the M240 (the new M60 designator for a 7.62 automatic weapon; think Rambo), and the heavy on-the-up guns like the .50 cal.

Each size weighs a bit more but also increases the effective range and battery life, so pick one appropriate for the equipment, and go get some.

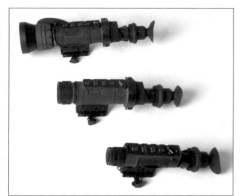

Collection of PAS-13s, from top to bottom: heavy, medium, and light ⩢

There are even smaller thermal sights on the market, and some that work just like the UNS discussed earlier. Insight Technologies has created the very small and light CNVD-T clip-on thermal weapons sight, which can be used as a stand-alone day or night sight, or can be mounted in front of your day scope without affecting the day scope's point of aim or impact.

For snipers, having one of these small units tucked into the drag bag is invaluable not only for identifying targets in adverse light and weather conditions, but for also reaching out and touching them.

The future is going to bring even more incredible technology to shooters on the front lines. Keep your eyes open, because we are all going to witness amazing things.

Imagine an auto-ranging 2x-40x lightweight sight with overlaid thermal and night vision capabilities, built-in visible and IR lasers along with IR floodlights and overt white light, integrated ballistics computers that calculate wind, elevation, angle, and environmental factors, and are attached to a lightweight, mostly carbon-fiber collapsible weapon that fires smart bullets—bullets that turn corners, track on laser-designated targets, and offer a user-selected terminal ballistic effect.

Think penetration, or explosion, or both.

Hey, a kid's got to dream, right?

We spent about eight weeks in the Philippines conducting a Foreign Internal Defense (FID) mission, training a couple of platoons of Filipino SEALs.

These are great missions, on more than one level. Living on the beach of beautiful Subic Bay with no other U.S. military around, for starters. The diving in the bay is amazing—warm water and a plethora of scuttled U.S. Navy ships at reachable depths left behind after the United States pulled out of the P.I. in World War II. A casino down the street, amazing tropical weather, and great cheap food.

Liberty benefits aside, the training we put on for the Filipino SEALs is top notch. They have modeled their SEALs after our program, but they just don't have access to the level of instruction we do, nor the equipment, training facilities, or food. (You should see how thin these guys were.) These guys are hard, though, and the majority had seen plenty of combat in the jungle of the Philippines chasing Abu Sayyaf and other terrorist organizations around.

Toward the end of our stay, we had them do a Final Training Exercise (FTX), a diverse mission profile that would take several days and force them to use the majority of their skills.

The mission was to send in a small Surveillance and Reconnaissance team to watch a village out in the jungle for a couple of days, send back detailed imagery of activity taking place there, and identify high-value targets (HVTs).

Actually, we were the high-value targets.

The recon team inserted by boat and did a long swim to a remote beach, then slipped up into the hinterland and worked their way just below a high ridge some distance off from the village. There were several of us glassing the jungle for movement for hours, doing our best to locate and bust the team coming in—and having no luck.

The simple fact was that these guys knew how to move in the jungle even better than we did. Most had grown up in it.

We had given up trying to find the team with night vision and binos—but we did happen to have a Raytheon PAS-13 thermal sight with us.

We flipped it on and started scanning the ridge—and in less than ten seconds,

⩘ **Glen selling his soul for a cheesy photo op in the Philippines. This was one of our little makeshift huts in the village where the Filipino SEALs executed a direct action assault.**

there they were: We picked up four perfect images of our boys up there, some 800 meters off.

If ever there was a proof of concept for technology, this was it!

10

LASERS

Military use of lasers is growing in the urban warfare theaters around the world. This technology offers a unique sighting and aiming capacity that can also have a tremendous psychological effect on enemy combatants.

Laser sights were first made practical with the invention of the laser diode in 1962. Prior to that, the power requirements and general fragility kept lasers confined to larger deployments (trucks, tanks, aircraft, ships, etc.). With the advent of the semiconductor laser in 1968 and subsequent miniaturization, the laser sight moved from science fiction to reality. (Not exactly a death ray, but then, we don't have flying saucers yet, either.)

The first laser sights were red in color, poorly collimated (meaning the laser beam would break up after traveling a short distance), and ate batteries like crazy. The beam was nearly round, in most cases, and extremely bright (laser sights are several times more intense than the sun at their operating color and frequency). Today they come in red, green, blue, and infrared, that last useful for aiming with night vision.

The first infrared solid-state lasers were highly collimated and had low power requirements. First used in laser printers, these devices quickly proved an excellent crossover technology into the military domain. With night vision scarce on the battlefield, their application was not very practical at first, but over time the IR laser targeting illuminator became the go-to choice for an "invisible" laser weapons sight.

Laser technology is used to range, illuminate, target, and mark targets for following on close air support (CAS) missions. Infrared laser systems add to our ability in certain environments to *own the night*.

Visible lasers, while widely known to the civilian world as a staple of Hollywood sci-fi, do in fact have some use in combat. They offer a unique sighting and aiming application, and can also be used as effective psychological warfare in situations that might not have escalated yet to the point of no return.

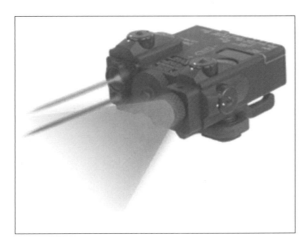

≈ **Laser Devices Inc.'s DBAL-A2 is a dual-beam aiming laser with both IR and visible laser pointers in one unit, combined with a focusable infrared laser illuminator; think giant flashlight when on night vision. The laser targeting is fully adjustable for windage and elevation. This unit is not designed to go out to extreme distances, but in close-quarter urban environments, enemy beware.**

I saw how effective visible laser can be while conducting sniper overwatch on the stricken U.S.S. *Cole* in October 2000, after suicide bombers killed over a dozen crew and nearly sank the ship as it restocked in the Gulf of Aden off Yemen.

I was set up on the bridge and built a hasty hide to blend in with the ship's superstructure. I had a .50 caliber and three LAW rockets, and orders to shoot to kill anything or person that compromised the perimeter. Not bad ROE, if you ask me.

The situation in Aden was tense and we were surrounded by hostiles who had major weaponry trained on the ship.

At night I shifted positions and started using my high-powered visible laser to scare the Yemeni weapons crews. After a few sessions of complimentary Lasik surgery, all crews shifted their weapons off the *Cole*. This brought a smile to my face and made the situation a little less tense—for us, anyway.

I don't care who you are, no one likes guns pointed at them. No matter who you are or where you are in the world, seeing an illuminated dot holding on your chest will cause you to rethink any mischief you had in mind. There have been many occasions where hidden snipers have used visible lasers to disperse volatile crowds or send a clutch of angry, poorly armed youths home to their mothers.

≋ The Insight M6X Laser Tactical Illuminator flashlight shown mounted on the Picatinny rail system. This one also shows the touch pad activation switch on the vertical grip, which makes for easy on, easy off. During the twentieth century, you would have had two separate units to accomplish what this much smaller single system can do.

One visible laser system commonly used in combat today is the M6X combination tactical flashlight and fully boresightable visible laser, manufactured by Insight and Streamlight (among others). These combination units can be used on pretty much anything from your pistol to your squad automatic weapon (SAW).

The M6X is a tiny light that can be manufactured specifically to mount directly to the frame of your particular sidearm; for use with rifles, they are attached to the weapon's 1913 rail system. Easy on, easy off with the throw of one locking lever—or, for more permanent attachment, there are some more fixed options that involve an Allen wrench or screwdriver at most.

The SOFLAM AN/PEQ-1 can be used for ranging and target designation. It is equipped with a 10x scope and mounting rails, and weighs about twelve pounds, depending on what you have attached to it. It can be used with or without tripod and can

≈ **A U.S. SF operator uses the SOFLAM to mark a target for air strike in Afghanistan in 2001.** (*U.S. Army*)

mark a target at distances of up to 20,000 meters (nearly 12.5 miles), though you'll want it on the tripod for distances in that range. It can also take a hard bounce without damage and is pretty easy to carry.

The main spot on a target at 5 kilometers is only about 2.3 meters square, so the precision of any ordnance is within that target square. That's why laser-guided bombs sometimes come through the window.

The SOFLAM AN/PEQ-1 was originally manufactured by Litton but, like most laser products, is now being manufactured by several leading companies. Either way, it is widely used and well known on the battlefield, and provides politicians with the ability to get SIGINT (signals intelligence) regarding the effectiveness of the strike along with actual and collateral damage.

The laser inside is a Nd:YAG (neodymium: yttrium alumina garnet) laser, which is to say that it's a solid laser that uses a laser diode to start the lasing of the main rod of Nd:YAG to create the beam. The beam is fairly efficient and the NiCad battery that powers the SOFLAM AN/PEQ-1 lasts a reasonable length of time. To add in an even more cool factor, the unit can be operated remotely, so that you can leave it behind, hidden, during a covert operation, to be activated at your leisure from a safe location.

One of the modern sniper's most effective tools is intimidation. When the enemy knows the sniper is out there somewhere, but cannot identify exactly where, it can be psychologically devastating.

In fact, history has shown again and again that well-trained snipers can crush the morale of entire armies.

There's nothing like a brightly visible laser spot on a bad guy's forehead to make him—and everyone around him—reevaluate the safety of his current location. Or even start thinking maybe it'd be smarter to be at home drinking tea with Mom.

11

GEAR

What gear you select to bring with you in the field depends completely on what the mission is. Are you doing sniper overwatch for a direct action mission? Or reconnaissance and surveillance, and if so, for how long, and in what kind of environment? Urban? Desert?

However, there are some items you'll be bringing with you on any sniper mission. To use militaryspeak, your kit is really broken down into three lines, called first-line, second-line, and third-line gear.

First-line gear is what is attached to your body, from your boots to your belt and whatever you can jam into your pockets. Let's break this out, head to toe.

When it comes to choosing outerwear and whatever additional gear you might bring with you, we have a universal expression: "Travel light, freeze at night." Sure, it may be 90°F or hotter during the day in Iraq, but it is the desert, and if you end up staying the night, you'd better be aware that the temp just might get down into the thirties.

↑ A SEAL sniper becomes part of the trees. Note the light reflection difference between the camouflage and the environment.

↑ A SEAL sniper conducts helicopter operations with an MK-12 rifle. A nylon cord (just past the bipod) provides reflexive support, making an extremely difficult shot more manageable.

Your head is the most important part of your body; best to keep it solidly above and between your shoulders. For you white boys operating in the desert, remember that shade is your friend, and it's also nice to have something that will cover your scope's eyepiece so sunlight doesn't reflect off it and blind you. In a word, we're talking here about a *hat*.

Floppy hats, baseball hats, ghillie hats for certain operations . . . these all work. When the sun goes down, though, it's best to have a black beanie tucked into your cargo pocket in your first-line gear, or at the least, in your second- or third-line gear. You won't regret it.

Sunglasses are key, too, and don't skimp on quality. Your vision depends on it, and as a sniper, maintaining that 20/20 and keeping your eyesight as sharp as possible isn't an afterthought, it is vital.

There's been a push by the big military for ballistic-quality lenses in their troops' protective eyewear. This is a good thing, but it may not be necessary or even practical

↑ Note that the lid of Glen's hat just covers the edge of the optical eyepiece to prevent glare.

⩲ **Digital camouflage patterns have been proven to increase not only the time to detection of an object, but also the recognition of the object as well.**

for every mission. A lot of mainstream eyewear companies are starting to release great lenses that offer fantastic optical quality in conjunction with ballistic safety.

Since we're working head to toe, we would be remiss if we didn't mention that you might be in a place for a long time, and it's not like you can (or should) fire up a cigarette. Chewing gum, on the other hand, is another story. During the second Gulf War, on the push from the Kuwait border toward Baghdad, the authors personally witnessed a teammate chew one piece of Trident spearmint gum for twenty-one straight days. *One piece.*

Camouflage clothing is constantly changing, as it should. The new MultiCam that the majority of the U.S. military is using is a big leap forward in the technology involved in camouflage and our understanding of how the human eye views things. Camouflage, too, is mission-dependent, but the new colors blended into the digital camouflage pattern work very well in a wide variety of environments.

Vegging up. The importance of properly selecting flora appropriate to your stalk cannot be overstated. ⩢

≈ **Brandon demonstrates concealment. Note the material covering the optics to prevent reflection and the proper use of vegetation.**

≈ **Cameras have replaced field sketching. Taking high-resolution photos of a target area and comparing the photos can reveal the movement of enemy snipers.**

Other nice touches are being built into today's uniforms, many of them adopted and adapted from ways that SpecOps guys were custom-modifying their uniforms fifteen years ago. The pockets that are now featured on both shoulders are great, and you'll also notice they are canted forward for easy access.

One problem guys operating in covert environments have to deal with, though, is Velcro. If you need something out of your Velcro'ed-shut pocket but have to maintain an extremely quiet profile, you could be in trouble. Sometimes guys will swap the Velcro out for buttons when they know they are heading into environments where noise discipline is paramount.

There are two canted pockets on the chest, but the pockets that used to be below those have been removed for operators, since we are usually tucking our cammie tops into our pants. In the old days, guys would just remove them. A pocket isn't any good when it's tucked into your crotch, and the extra material is annoying.

A good rigger's belt is nice: It's secure and it can be used in a hoist operation or, in a pinch, to rappel.

Gloves are always important and essential, if you're doing a long stalk through dicey terrain. Lots of guys are using gloves designed for professional baseball or football players, but there are also new tactical gloves being sold that offer the best of most worlds. They may have a carbon fiber protective plate across the back of the knuckles, some extra protection on the palm, and well-designed fingers for increased sensitivity and dexterity.

Footwear is a personal choice; there really is no right boot or shoe for sniper ops. Again, it's mission-dependent.

In one leg pocket you should have some sort of medical blowout kit for self-aid that includes a tourniquet and some blowout bandages.

Somewhere in your first line you should have your blood chit. A blood chit is printed on waterproof, tear-proof paper, with its four corners perforated with tear-off pieces with a serial number imprinted on them that matches the main chit. In an emergency, these tear-off pieces can be given to anyone you deem trustworthy enough to assist you, and can be turned in for a reward at a U.S. government installation.

Every chit's ID number is logged and associated with that individual, so if a corner does get turned in, we will know exactly who is on the run and that they are alive, or at least that they were alive recently enough to hand off this piece of chit.

In addition to the large American flag printed on the chit's front face, there is also a block of text, printed in whatever different languages are relevant to the region you are operating in, that says something to this effect:

I am a citizen of the United States of America. I do not speak your language. Misfortune forces me to seek your assistance in obtaining food, shelter, and protection. Please take me to someone who will provide for my safety and see that I am returned to my people. My government will reward you.

In another pocket you want to pack your Department of Defense common access card (DOD CAC) iden-

⌃ **SEAL snipers in Afghanistan engaging the enemy. Assorted gear.**

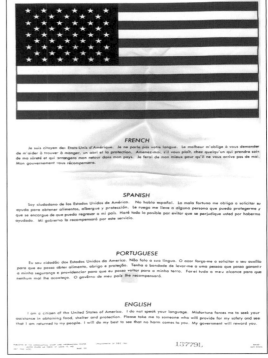

⌃ **An example of an old bloodchit**

⌃ A chest rig (worn here by Glen) can carry a lot of kit comfortably. Peltor earmuff headset works great for limiting eardrum-shattering noise and accentuating other sounds.

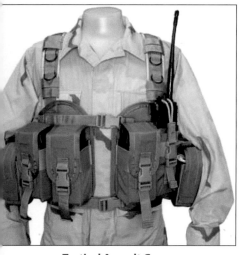

⌃ Tactical Assault Gear (TAG); Front Chest Rig

tification and a decent amount of cash (in both U.S. and local currency). If the shit hits the fan and you are on your escape and evasion (E&E) plan, cash may be the only item you need to survive. "Money talks" is a universal expression that you'll find applies anywhere on the globe.

And if you are on a covert operation and not be wearing a uniform, that ID will likely be the only thing that gets you back into fortified friendly zones.

A small survival kit should take up your other cargo pocket space. Something to eat (even if it's just a Clif Bar), some water purification tablets, a good pocket knife, a signal mirror, some cammie paint, a good lighter . . . you get the idea. It's also nice to have a small, hand-held GPS and perhaps a map in case you get in trouble and have to call in an emergency CAS mission to save your ass or to take out a significant target.

Your second-line gear is basically your fighting kit. This will include your primary weapon and perhaps your secondary, as well, if you have a pistol slid into a pocket somewhere. Most likely you'll have some type of load-bearing equipment (LBE), too.

A Rhodesian chest rig or similar (also referred to as a slick rig) is popular among snipers, as it keeps your hips free for ease of movement, and you can carry pretty much everything you might need for the op.

Your second line of kit will also have an IR/overt strobe light for marking your position to friendlies, night vision, extra batteries for optics or night-vision goggles, water, extra ammunition, binos, a range finder (if there isn't one built into your binos), and perhaps some 550 cord/bungee cords/zip ties/rubber bands/paperclips/thumbtacks to help set up a shooting position or hide site.

Some sort of comms will be stashed in your rig, too, whether it's a handheld radio, cell phone, or other device. Communications can often be your most powerful weapon. A sniper can feel well endowed when in a strong shooting position with a .50-cal rifle loaded and ready—but controlling a flight of F18s for CAS strikes is truly powerful and provides quite the edge in firepower.

Third-line gear is equipment you might carry in a backpack or drag bag. The variety of items that might be included here could fill another book; it entirely depends on what you've been tasked with.

A digital SLR camera with a telephoto lens might be part of the kit, with a ruggedized laptop and satellite uplink for sending photos or video anywhere in the world. The powers that be enjoy being kept in the loop when a small team is in the field, and with that equipment you can provide detailed imagery customized for your audience.

A sniper in a forward observation position could identify and photograph primary breach points on a target, patrol patterns, enemy numbers, equipment, and uniforms. This is pertinent and powerful information for follow-on missions.

If you and your team know you are going to be in the field for an extended stay, your third line can and will grow to absurd sizes. It's not unusual to see someone humping with a 45-kilogram ruck or more. Hide site material,

≽ This drag bag by RS Tactical can be worn as a backpack when your sniper rifle cannot (and should not) be used as your primary weapon for long infils or for close quarters until in your final shooting position.

≽ A SEAL sniper trains with a sniper variant of the FN MK20 SCAR-Heavy.

digging equipment, enough batteries to power all the electronic equipment, antennae, food, and the most important and heaviest of them all, *water.*

The most prominent conflicts the United States is currently engaged in are in desert climates. Water is critical, especially for the shooter. From a physiological standpoint, the first thing that starts degrading when dehydration starts setting is your vision. Stay hydrated and keep that clear sight picture.

12

THE MORALITY OF KILLING

Too often we get asked what it's like to take another human being's life.

This is a very difficult and uncomfortable question to answer and we encourage you not to ask it to anyone unless you share a close relationship with that person. However, we did think it's important enough an issue to at least address the topic of killing and give you our perspective.

What are morals? They involve culturally accepted standards of what is right and what is wrong. But the only thing certain here is that there exists *no* universally accepted set of morals.

In fact, the particulars of what is moral and what is not vary widely from culture to culture. What is right in some countries is wrong in others. For instance, in some Middle Eastern countries it is still acceptable for a man to kill his wife.

Approaching the whole question of the morality of killing is far from simple or absolute.

In the Special Operations community, we have a common belief that there are three types of people in the world: wolves, sheepdogs, and sheep.

The wolves are what most would consider evil people. They are the rapists and murderers of the world, preying on the weak and using violence and others' fear to achieve their goals.

Then there are the sheepdogs. They can look similar to the wolves, in many ways, and can sometimes even be mistaken for them fairly easily, but have a mission that is virtually opposite to that of the wolves. Rather than preying on others, they are there to protect those who cannot protect themselves. They exist solely to look after the safety of the flock.

Finally, there are the sheep: everyday good people who go about their lives, able to do so in safety only because they are protected from the wolves. They don't necessarily like to be around or acknowledge the sheepdogs, because this can make them nervous and more uncomfortably aware that there really are wolves out there ready to cause them harm.

The Special Operations personnel of the world are sheepdogs.

Former SEAL Scott Tyler illustrates this view in this recollection from his days in sniper school:

My fondest memories of sniper school include an introduction by the SEAL Lead Petty Officer instructor who was proctoring our class. As I sat in my seat what I heard him say was, "If you don't lay in bed at night, hoping and praying that someone breaks into your house so that you can shoot them and kill them, you don't belong in this class."

≈ Despite the benevolence of the major world religions' basic tenets, killing in the name of a God is both a historical and a contemporary fact.

To folks who don't have any exposure to the tradition of arms, this may seem like a horrific and offensive approach to something as sacred as taking the life of another human being. The reality is that there are monsters in the world who prey upon others, who will take as much as they can unless someone is willing to step forward and stop them. There are also people who live in fear of those monsters— and then there are the people who embrace the reality of what it means to stand up and take the fight directly to

≈ SEALs awaiting extraction by an MH-53 PAVE LOW helicopter

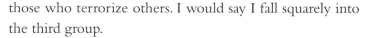

those who terrorize others. I would say I fall squarely into the third group.

What that LPO was really saying is more about being in that third group, and wanting to be in situations where you are forced to take appropriate action. No one I worked with wanted to go out and kill people just because they could. We were all motivated to be the people trusted with the skill, intuition, and judgment to make the right call at the right time.

When you are in a hide site, isolated from any support, in the middle of bad-guy land, and all you have is your gun and your shooting partner, you want to be next to someone who is not horrified at the prospect of having to take a life when the moment comes. Because it *will* come, and your survival depends on unflinching determination and resolve, enabled by training, to eliminate the enemy before they eliminate you.

As Winston Churchill said, "We sleep soundly in our beds because rough men stand ready in the night to visit violence on those who would do us harm."

From our personal experience and extensive conversations with other operators and snipers, we've found that everyone deals with death and killing differently.

Much of this difference is a reflection of the unique individual, their upbringing, and the situation that brought them to the killing fields.

SEALs conduct fast-rope training in Southeast Asia. ≫

While the snipers of Stalingrad were fighting for their very lives and for that of their families and their homeland, the German snipers they faced had decidedly different motivations.

Major Hesketh-Pritchard, the legendary World War I–era British sniper, sums it up well in Andy Dougan's *Through the Crosshairs*:

> The sniper must be able to calmly and deliberately kill targets that may not pose an immediate threat to him. It is much easier to kill in self-defense or in the defense of others than it is to kill without apparent provocation. The sniper must not be susceptible to emotions such as anxiety or remorse. Candidates whose motivation towards sniper training rests mainly in the desire for prestige may not be capable of the cold rationality the sniper's job requires.

The U.S.S. *Cole* is an awful reminder to remain vigilant. Bad people are out there and are full of hate toward those who believe differently than they do. (*U.S. Navy*) ≽

During SEAL sniper training, you learn a system and become incredibly proficient at the mental and mechanical side of making a long-range shot at a silhouette target. We had several celebrity coaches and trainers guest-teaching in our three-and-half-month-long course.

One of these men discussed at length the mental aspect and potential repercussions of taking someone's life. He was a decorated combat veteran, with over fifty confirmed kills during multiple deployments in Vietnam, who later became a champion long-range competition shooter. He was also a religious man.

Here's what he told our students:

When I was over there, in the bush, watching and waiting for my next target, you had a fair amount of time to reflect on your life, and that of the man whose life you were going to attempt to take.

All gave, some gave all. U.S. Special Operations Memorial. ⌄

I figured I was trained to do a job, just like my VC counterpart, and if and when that person took shape in my crosshairs, well then, it was just his time, and God was calling him home.

I also realized that on any given day it could be my turn in someone else's crosshairs, and I held no ill will toward the man who held that rifle, as he was just doing his job, too, and it was just my time to go home.

It was a powerful message delivered by a humble man, and it resonated with everyone in the class, from the atheist to the practicing Christian.

The one thing we can tell you is that 99 percent of the people we know who are Special Operations snipers have no problem rationalizing the kill.

We *have* met some guys who have a problem with it; these guys ultimately get transitioned out of the Special Operations community into roles for which they are more suited. In our community, the training is long and intense, and you would have to be pretty obtuse to not reflect on what it is you are being trained to do. Those people usually don't make it to our community; those who do have already rationalized their job in their minds.

However, thinking about shooting an enemy combatant during training and experiencing your first genuine combat are two very different beasts, indeed, and there will almost certainly be a moment when, despite all the training you've gone through, you take a deep breath and strip it all down to the bare truth: You are a man, recognizing another man who is now dead because of your actions.

This the moment of truth is the point when individual rationalizations start to take hold, and your next sleep will either be as peaceful as a child's or completely disturbed.

In Sasser and Roberts's *One Shot—One Kill*, a marine captain and legendary Vietnam–era sniper and instructor reflects on the type of mind-set a man must have to be a great sniper:

There is no hate of the enemy. Psychologically, the only motive that will sustain the sniper is knowing he is doing a necessary job and having the confidence that he is the best person to be doing it. When you look through a scope the first thing you see is the eyes. There is a lot of difference between shooting at an outline . . . and shooting at a pair of eyes. Many men can't do it at that point.

As recounted in Dougan's *Through the Crosshairs*, Carlos Hathcock sums up his thoughts about being a sniper in his characteristically simple and still-relevant way:

I like shooting, and I love hunting. But I never did enjoy killing anybody. It's my job. If I don't get those bastards, then they are gonna kill a lot of those kids dressed up like Marines. That's the way I look at it. . . . But I never went on any mission with anything in mind other than winning this war and keeping those . . . bastards from killing more Americans.

THE FUTURE OF SPECIAL OPERATIONS: SNIPING INTO THE TWENTY-FIRST CENTURY

It is both amazing and frightening that neither the U.S. military nor state and federal law enforcement communities have standardized sniper programs in place.

I'm not saying there aren't great programs out there. The U.S. Navy SEAL sniper program is one of the best in the world. What I'm pointing out is that the U.S. sniper community needs to get it together and standardize training and methodology.

For the most part, the military is getting closer in this regard, but I'm shocked at the lack of standardization and low training standards in the American law enforcement communities. I've worked with some of the biggest and most well-funded law enforcement SWAT sniper programs and have been shocked at the lack of training these solid operators receive. I have yet to run across a law enforcement sniper program that has incorporated the best practices of the sniper community.

⌃ This is a great example of modern advances in nanotech camouflage technology. Note the person's camo jacket blending in with the surrounding environment.

A small SEAL element is delivered to a submarine via chopper. ⌄

To be fair, military snipers, unlike those in civilian law enforcement, have the luxury of the battlefield environment and favorable rules of engagement. In most cases we aren't concerned with hostages or other casualties involved in the shots we take.

My SEAL friend Jason recently returned from Afghanistan, where his unit killed close to two hundred bad guys. On one op he said he got two guys with a single shot: His round hit the driver, passed clear through him, and then hit and killed the passenger who was inline (we call this bullet path). Fortunately for Jason, the passenger up front was someone who needed killing, too—not some hostage of other innocent victim of what is benignly called "collateral damage."

Law enforcement has different rules of engagement, and the stakes are usually much higher because citizen hostages are involved, creating a critical situation where there is zero room for error.

Not that military snipers *never* see this sort of situation. You probably remember the April 2009 incident when three Somali pirates were shot in the rescue of Captain Richard Phillips of the *Maersk Alabama*. Three SEAL snipers shot those pirates in complete darkness, shooting from one moving platform onto another moving platform, with flawlessly coordinated surgical precision—and it was a good thing, too, because one poorly executed shot could have executed the man they were trying to save.

But there *was* no poorly executed shot.

Those three simultaneous kill shots expertly diffused a volatile international situation and directly resulted in Phillips's rescue. These snipers were extremely well trained and they accounted for every factor, both human and environmental, guaranteeing the mission's success.

I've found that most local law enforcement agencies allow ego and rivalries to get in the way of solid training, and for that as well as other reasons, their sniper programs are generally severely lacking. I've been on the range with qualified snipers who did not even grasp the fundamentals of internal, external, or terminal ballistics. Many (even most) argued with me for wanting to push their training past the 100-meter threshold.

I recognize that the average police sniper shot is in the neighborhood of 72 meters, but it is also extremely important to train *beyond* your typical practical application. If you can consistently drive nails at 540 meters under adverse conditions—artificially induced stress, wind, barometric pressure, high angle, etc.—then the shot you have to make when the call comes down will be like walking the dog.

I am baffled when people argue with me about the reasoning behind this methodology. If you look at every world-class athletic program, you'll find they all train for conditions much harder then what they will actually face in competition.

It is my hope that the U.S. law enforcement sniper community realizes the importance of standardizing their training and how important the role of the sniper is when the time comes.

One thing is for sure: The expert sniper, along with the critical role he or she plays, is here to stay. As the threat of terrorism increases, I see an increase in employment for snipers going well into the twenty-first century in both the military and law enforcement communities.

It's not that we live in a more dangerous world, but that as we become more of a global society, the threat to society is much more asymmetric in nature. In fact, globalization is threatening more traditional societies, which in turn often look to combat the threat they perceive coming from the forces of modernization with terrorism. In many cases, it is the only way they have (or at least, the only way they can *see*) to effectively engage more affluent societies.

Know your enemy. Taken of enemy KIA in Afghanistan. ⩗

As this trend increases, developed nations will have to adapt in order to deal with this modern asymmetric threat. And a critical part of this adaptation is the increased focus on and use of unconventional strategy and tactics—a toolbox whose central tool is the twenty-first-century sniper.

While the actual weapons and technology employed by the modern sniper will continue to change rapidly, the mission will remain the same: delivering precise solutions with zero collateral damage and maximum effectiveness: a stealthy and precise kill shot such that the target is taken completely by surprise and is dead before he even hits the ground.

Radical changes in technology are occurring that the sniper community can leverage to their great advantage. We now have technology in optics that fuses together the IR and thermal spectrum, nanotech camouflage that bends light, and lasers coupled to sniper scopes that are capable of precision wind calculations.

"Americans sleep peacefully at night only because rough men stand ready to do violence on their behalf."—U.S. Navy SEALs »

It's exciting to see this new technology and how it can increase the modern-day sniper's capacity and effectiveness on the job.

The problem, though, has never been with innovation; the biggest problem has always been with exercising the organization wisdom and political will to put the most current technology into the operators' hands. Unfortunately, the bureaucracy involved with government acquisition too often gets in the way. Typically, large systems integrators for the government deliver untimely and irrelevant solutions.

☆ **Every SEAL sniper is qualified on the Drager rebreathing system. It provides an excellent insertion capability.**

I once was advising a large, billion-dollar defense company that was working on a computer system for snipers. When I saw what they were doing, I recognized immediately that what they

« **A SEAL performing a High Altitude, High Opening (HAHO) training jump**

⌃ **Desert training**

had been developing for years, at the cost of tens of millions of U.S. taxpayer dollars, I could buy off the shelf for thousands.

They were so focused on their program that they were blind to this fact. And even after coming to terms with the reality of the situation, the program director told me he was going forward anyway—because without the program, his job would go away!

Complete madness, and a tremendous waste of taxpayer dollars—dollars that could be put to far better use in equipping our frontline snipers with cutting-edge technologies.

The fact is, the U.S. government acquisition system is broken, and we need to fix it—or our enemies will start being better equipped than we are.

As dire as that sounds, the picture is not totally bleak. I have personally been involved in this process on the military side, and I can attest that Admiral Olson of U.S. Special Operations Command (U.S. SOCOM) has recognized that this is a problem and SOCOM is in the process of fixing it as you read these words.

IF YOU CAN'T AFFORD A BALLISTIC COMPUTER, OR EVEN IF YOU CAN

Knowing what your rifle does with the ammunition you choose (or that you are issued) and how it performs under different conditions is paramount to making accurate shots, especially if you adhere to the "one shot, one kill" rule.

Keeping accurate logs of your shooting is a big help in knowing how you and your rifle will perform under a variety of conditions. Most shooters call these logs "dope" books.

These will start falling by the wayside as ballistic software becomes more prevalent in the world, but it's important to embrace the old before learning the new. Batteries die and electromagnetic pulses fry electronics—so keep your round count and dope books up to date.

This following data charts were provided by Dave Durham at CheyTac.

Sniper Chart 1: Barrel Log

Barrel Log

Date	Ammunition	Lot	#Fired	Total	Date	Ammunition	Lot	#Fired	Total

This Page

Previous Page

Total

Notes:

Sniper Chart 2: Come Ups

TEMP RANGE: _____ deg F

LOAD DATA:
Bullet: _____
Velocity: _____
Altitude: _____

NOTES:
_____
_____
_____
_____
_____

Come-Up	From 0	Come-Up	From 0	10 MPH
100		150		
200		250		
300		350		
400		450		
500		550		
600		650		
700		750		
800		850		
900		950		
1000		1050		

Sniper Chart 3: Formulas

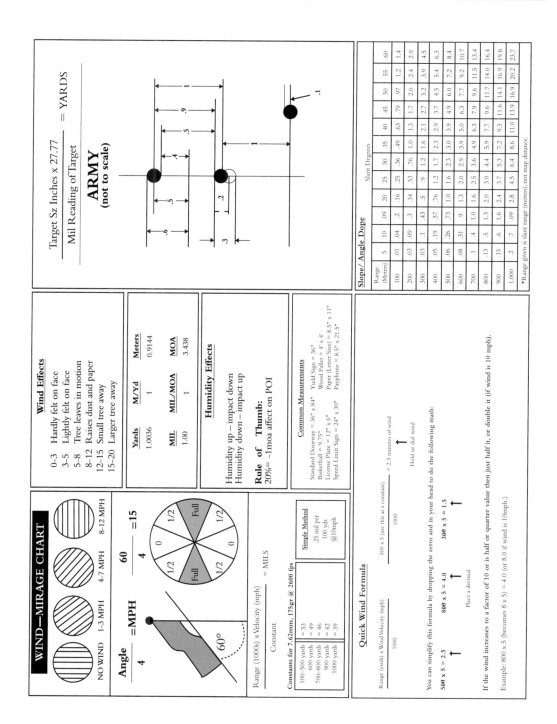

WIND—MIRAGE CHART

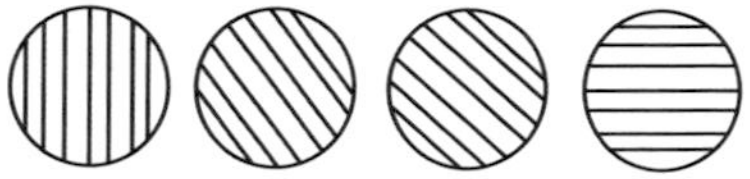
NO WIND
1-3 MPH
4-7 MPH
8-12 MPH

$$\frac{\text{Angle}}{4} = \text{MPH} \qquad \frac{60}{4} = 15$$

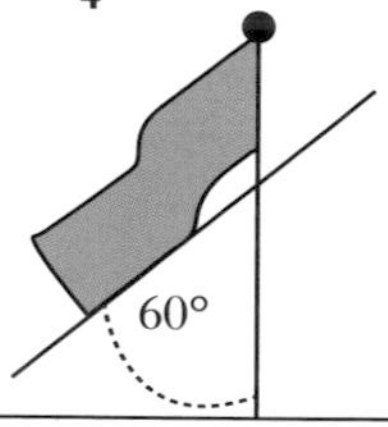
60°

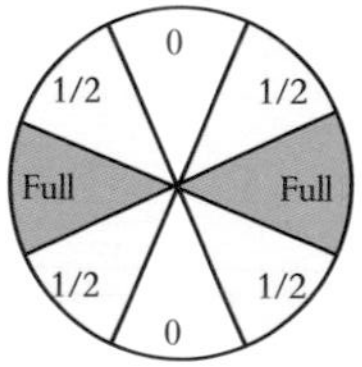
0
1/2
1/2
Full
Full
1/2
1/2
0

$$\frac{\text{Range (1000s) x Velocity (mph)}}{\text{Constant}} = \text{MILS}$$

Constants for 7.62mm, 175gr @ 2600 fps

100–500 yards	= 53
600 yards	= 49
700–800 yards	= 46
900 yards	= 42
1000 yards	= 39

Simple Method

.25 mil per
100 yds
@10mph

Wind Effects

0–3	Hardly felt on face
3–5	Lightly felt on face
5–8	Tree leaves in motion
8–12	Raises dust and paper
12–15	Small tree away
15–20	Larger tree away

Yards	M/Yd	Meters
1.0036	1	0.9144

MIL	MIL/MOA	MOA
1.00	1	3.438

Humidity Effects

Humidity up – impact down
Humidity down – impact up

Rule of Thumb:
20%= -1moa affect on POI

Common Measurements

Standard Doorway = 36" x 84"
Basketball = 9.75"
License Plate = 12" x 6"
Speed Limit Sign = 24" x 30"
Yield Sign = 36"
Wood Pallet = 4' x 4'
Paper (Letter Size) = 8.5" x 11"
Payphone = 8.5" x 21.5"

$$\frac{\text{Target Sz Inches x 27.77}}{\text{Mil Reading of Target}} = \text{YARDS}$$

ARMY
(not to scale)

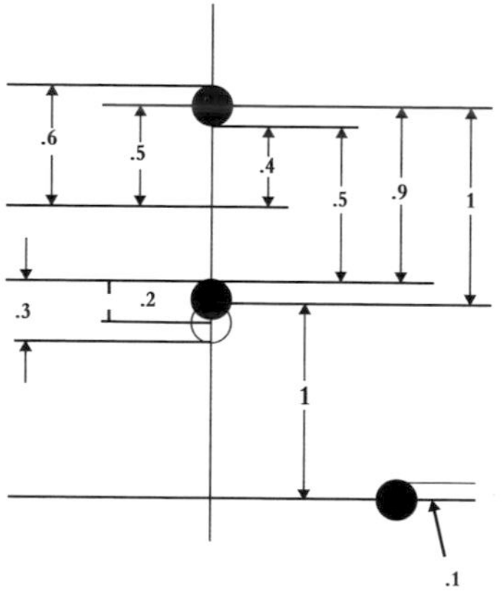
.6
.5
.4
.5
.9
1
.3
.2
1
.1

Quick Wind Formula

$$\frac{\text{Range (yards) x Wind Velocity (mph)}}{1000} \qquad \frac{\text{500 x 5 (use this as a constant)}}{1000} = \text{2.5 minutes of wind}$$

↑ Hold or dial wind

You can simplify this formula by dropping the zeros and in your head to do the following math:

5~~00~~ x 5 = 2.5 ↑ 8~~00~~ x 5 = 4.0 ↑ 3~~00~~ x 5 = 1.5 ↑

Place a decimal

If the wind increases to a factor of 10 or is half or quarter value then just half it, or double it (if wind is 10 mph).

Example: 800 x 5 (becomes 8 x 5) = 4.0 (or 8.0 if wind is 10mph.)

Slope/ Angle Dope

Range (Meters)	Slant Degrees 5	10	.09	20	25	30	35	40	45	50	55	60
100	.01	.04	.2	.16	.25	.36	.49	.63	.79	.97	1.2	1.4
200	.03	.09	.3	.34	.53	.76	1.0	1.3	1.7	2.0	2.4	2.9
300	.03	.1	.43	.5	.9	1.2	1.6	2.1	2.7	3.2	3.9	4.5
400	.05	.19	.57	.76	1.2	1.7	2.3	2.9	3.7	4.5	5.4	6.3
500	.06	.26	.73	1.0	1.6	2.3	3.0	3.9	4.9	6.0	7.2	8.4
600	.08	.31	.9	1.3	2.0	2.9	3.9	5.0	6.3	7.7	9.2	10.7
700	.1	.4	1.0	1.6	2.5	3.6	4.9	6.3	7.9	9.6	11.5	13.4
800	.13	.5	1.3	2.0	3.0	4.4	5.9	7.7	9.6	11.7	14.0	16.4
900	.15	.6	1.6	2.4	3.7	5.3	7.2	9.3	11.6	14.1	16.9	19.8
1,000	.2	.7	.09	2.8	4.5	6.4	8.6	11.0	13.9	16.9	20.2	23.7

*Range given is slant range (meters), not map distance.

Sniper Chart 4: Zero Data

Zero Data

Date	Ammunition	Distance

Rifle/scope

Time	Location	Temp

Altitude	Humidity	Baro. Press.

Wind

☐ Light (8) ☐ Mad (7) ☐ Heavy (15) ☐ _____

Light

☐ Bright ☐ Hazy ☐ Overcast ☐ Changing

Mirage

Shot #	1	2	3	4	5	6	7	8	9	10
Elev.										
Wind										
Call	☐	☐	☐	☐	☐	☐	☐	☐	☐	☐

Shot #	11	12	13	14	15	16	17	18	19	20
Elev.										
Wind										
Call	☐	☐	☐	☐	☐	☐	☐	☐	☐	☐

Notes:

Target Size =

Sniper Chart 5: Zero Summary

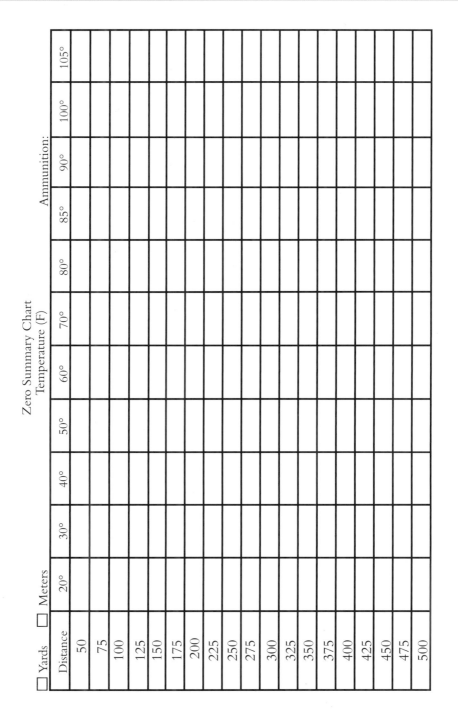

Zero Summary Chart
Temperature (F)

Ammunition:

☐ Yards   ☐ Meters

Distance	20°	30°	40°	50°	60°	70°	80°	85°	90°	100°	105°
50											
75											
100											
125											
150											
175											
200											
225											
250											
275											
300											
325											
350											
375											
400											
425											
450											
475											
500											

Ammunition: _____

Zero Summary Chart
Temperature (F)

☐ Yards ☐ Meters

Distance	20°	30°	40°	50°	60°	70°	80°	85°	90°	100°	105°
50											
75											
100											
125											
150											
175											
200											
225											
250											
275											
300											
325											
350											
375											
400											
425											
450											
475											
500											

VELOCITY COMPARISON TABLES

"I know what you're thinking, punk. You're thinking, 'Did he fire six shots or only five?'

"Now, to tell you the truth, I forgot myself in all this excitement. But being this as is a .44 Magnum, the most powerful handgun in the world, and would blow your head clean off, you've got to ask yourself one question: 'Do I feel lucky?' Well, do ya, punk?"

—*Dirty Harry*

It was 1971 when Clint Eastwood's Dirty Harry spoke those iconic lines. Today there are even more powerful handgun cartridges than the .44 Magnum available for big-bore pistol shooters.

What does a big, loud, high-recoil handgun have do to with snipers? Nothing, really. But a look at the following ballistic table will demonstrate just how powerful today's sniper rifles are.

Table 1: Velocity and Energy Comparisons

Cartridge BC	Bullet Wt	MV	ME	V @ 100	E @ 100	V @ 300	E @ 300	V @ 500	E @ 500	V @ 1,000	E @ 1,000
.44 Mag/0.157	240	1,760	1,650	1,361	986	951	482	784	328	528	148
.223 Rem/0.372	77	2,750	1,293	2,503	1,071	2,049	717	1,645	463	1,007	173
.308/0.458	168	2,600	2,521	2,410	2,166	2,055	1,574	1,733	1,120	1,142	486
.308/0.470	167	2,690	2,683	2,510	2,320	2,146	1,708	1,824	1,233	1,201	535
.300WM/0.533	190	2,950	3,671	2,773	3,243	2,439	2,509	2,129	1,912	1,460	899
.338/0.587	250	2,950	4,830	2,789	4,316	2,484	3,424	2,198	2,682	1,532	1,303
.338/0.675	250	2,969	4,892	2,828	4,438	2,559	3,634	2,305	2,949	1,739	1,678
.338/0.768	300	2,800	5,222	2,680	4,785	2,451	4,001	2,232	3,318	1,734	2,002
.408 CheyTac/1.0	419		7,700								
.50 BMG/1.05	750		11,200								

Table 2: Time to Target and Drop Comparison

(Note the drop for all distance is negative)

Cartridge	BC	Bullet weight	Time to 100	Drop @ 100	Time to 500	Drop @ 500	Time to 2,000	Drop @ 2,000
.44 Mag	0.157	240	0.1942 s	6.6 inches	1.4359 s	319.4 inches	12.99 s	20,496.6 inches
.223 Rem	0.372	77 M. King	0.1142 s	2.4 inches	0.7058 s	81.6 inches	5.69 s	4,480.8 inches
.308 Win	0.458	168 M. King	0.1198 s	2.7 inches	0.7073 s	84.6 inches	5.09 s	3,663.8 inches
.308 Win	0.47	167 Scenar	0.1156 s	2.5 inches	0.6777 s	78.1 inches	4.90 s	3,363.0 inches
.300 Win Mag	0.533	190 M. King	0.1049 s	2.1 inches	0.5988 s	62.2 inches	4.26 s	2,483.0 inches
.338 Lapua	0.587	250 M. King	0.1046 s	2.1 inches	0.5893 s	60.8 inches	4.05 s	2,229.6 inches
.338 Lapua	0.675	250 Scenar	0.1035 s	2.0 inches	0.5735 s	58.4 inches	3.64 s	1,857.4 inches
.338 Lapua	0.768	300 Scenar	0.1095 s	2.3 inches	0.6000 s	64.4 inches	3.62 s	1,887.1 inches

Table 1 compares the velocities and energies of the .44 Mag rifle cartridge to sniper cartridges. Table 2 compares the time to target and the drop from the bore of the .44 Mag to the same rounds.

Notes on Table 1

You can see that Harry's .44 Magnum develops 1,650 foot-pounds at the muzzle. The .308 at five hundred yards has almost as much energy, and at one thousand yards it still has one-third the energy of the .44.

Almost unbelievably, the .338 Lapua 250-grain Scenar and .300-grain MatchKing have more energy at one thousand yards than the .44 Mag at the muzzle.

Even the hottest hand load for the .44 Mag can generate a muzzle velocity of only about 1,400 fps.

Notes on Table 2

In this table you can see that the .44 Mag is not even close to the sniper cartridges in time to target (thirteen seconds to reach two thousand yards) and the drop is dramatic—a negative (i.e., downward) 20,496 inches at two thousand yards. For you math freaks, that's 1,705 feet, or 568 yards.

At two thousand yards, 1 MOA is twenty inches, so the drop would be:

$$20,496 \div 20$$

. . . or 1,024 MOA of elevation. (Can you say "impossible holdover"?)

Sniper rifles and their cartridges are simply awesome examples of man's ability to find a better way to fight his enemies.

Feeling lucky, Harry?

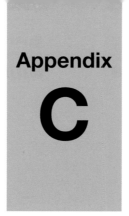

EXBAL BALLISTICS CHARTS

Black-Hills New Mfg: .223 Remington: Sierra MatchKing 0.224" 77gr HPBT

	SIGHT-IN DATA	FIELD DATA	POINT BLANK RANGE DATA			
			TARGET HEIGHT	SIGHT-IN DISTANCE	POINT-BLANK HIGH	RANGE LOW
Muzzle Velocity (fps)	2750	2750				
Bullet Weight (grains)	77					
Sight Height (in)	1.5		(in)	(yd)	(yd)	(yd)
Sight-in Distance (yd)	100		2	158	96	178
Altitude (ft)	0	0	4	198	114	227
Temperature (deg F)	59	59	6	228	127	264
Pressure @ Sea Level (in Hg)	29.53	29.53	8	254	139	297
Relative Humidity (pct)	78.0	78.0	10	276	153	321
Wind Velocity (mph)		0				
Wind Angle (degrees)		180 (6.0 O'Clock)				
Incline Angle (degrees)		0				
Moving Target Speed (mph)		0.0				
Ballistic Coefficient	0.372	0.362	0.362	0.343		
Lower Velocity Limit (fps)	3000	2500	1700	0		
MAX APPARENT TRAJECTORY (in)		0.1				

TARGET	SIGHT ADJUSTMENTS			TRAJECTORY VALUES					Drop from	ARRIVAL
DIST (yd)	ELEV MOA	WIND MOA	LEAD MOA	ELEV (in)	WIND (in)	LEAD (in)	VELOCITY (fps)	ENERGY (ft-lb)	bore line (in)	TIME (sec)
0	0.00	0.00	0.00	−1.5	0.00	0.00	2750	1293	0.0	0.0000
50	0.25	0.00	0.00	−0.1	0.00	0.00	2625	1178	−0.6	0.0558
100	0.00	0.00	0.00	−0.0	0.00	0.00	2503	1071	−2.4	0.1143
150	0.75	0.00	0.00	−1.3	0.00	0.00	2384	972	−5.6	0.1757
200	2.00	0.00	0.00	−4.1	0.00	0.00	2269	880	−10.4	0.2401
250	3.25	0.00	0.00	−8.5	0.00	0.00	2157	975	−16.8	0.3079
300	4.75	0.00	0.00	−14.9	0.00	0.00	2049	717	−25.2	0.3792
350	6.25	0.00	0.00	−23.3	0.00	0.00	1943	646	−35.5	0.4544
400	8.00	0.00	0.00	−34.0	0.00	0.00	1842	580	−48.2	0.5336
450	10.00	0.00	0.00	−47.3	0.00	0.00	1743	519	−63.5	0.6173
500	12.00	0.00	0.00	−63.5	0.00	0.00	1645	463	81.6	0.7058
550	14.50	0.00	0.00	−82.8	0.00	0.00	1551	411	−102.9	0.7997
600	16.75	0.00	0.00	−105.8	0.00	0.00	1462	366	−127.8	0.8992
650	19.50	0.00	0.00	−132.8	0.00	0.00	1379	325	−156.8	1.0048
700	22.50	0.00	0.00	−164.4	0.00	0.00	1303	290	−190.3	1.1167
750	25.50	0.00	0.00	−201.1	0.00	0.00	1235	261	−229.0	1.2350
800	29.00	0.00	0.00	−243.5	0.00	0.00	1174	236	−273.4	1.3596
850	32.75	0.00	0.00	−292.2	0.00	0.00	1122	215	−324.0	1.4903
900	37.00	0.00	0.00	−347.8	0.00	0.00	1079	199	−381.6	1.6267
950	41.25	0.00	0.00	−410.9	0.00	0.00	1041	185	−446.6	1.7683
1000	46.00	0.00	0.00	−481.9	0.00	0.00	1007	173	−519.7	1.9150
1050	51.00	0.00	0.00	−561.6	0.00	0.00	978	163	−601.3	2.0662
1100	56.50	0.00	0.00	−650.4	0.00	0.00	951	155	−692.0	2.2219
1150	62.25	0.00	0.00	−748.7	0.00	0.00	927	147	−792.3	2.3818
1200	68.25	0.00	0.00	−857.2	0.00	0.00	906	140	−902.8	2.5458
1250	47.50	0.00	0.00	−976.4	0.00	0.00	886	134	−1023.9	2.7136
1300	81.25	0.00	0.00	−1106.6	0.00	0.00	867	128	−1156.1	2.8852
1350	88.25	0.00	0.00	−1248.5	0.00	0.00	849	123	−1299.9	3.0606
1400	95.75	0.00	0.00	−1402.5	0.00	0.00	831	118	−1455.9	3.2398
1450	103.25	0.00	0.00	−1569.1	0.00	0.00	815	113	−1624.5	3.4229
1500	111.25	0.00	0.00	−1749.0	0.00	0.00	798	109	−1806.4	3.6098
1550	119.75	0.00	0.00	−1942.7	0.00	0.00	783	105	−2002.0	3.8005
1600	128.25	0.00	0.00	−2150.7	0.00	0.00	768	101	−2211.9	3.9952
1650	137.25	0.00	0.00	−2373.6	0.00	0.00	754	97	−2436.8	4.1937
1700	146.75	0.00	0.00	−2612.0	0.00	0.00	741	94	−2677.2	4.3961
1750	156.50	0.00	0.00	−2866.6	0.00	0.00	728	90	−2933.7	4.6025
1800	166.50	0.00	0.00	−3137.9	0.00	0.00	715	87	−3207.0	4.8128
1850	176.75	0.00	0.00	−3426.6	0.00	0.00	703	84	−3497.7	5.0271
1900	187.75	0.00	0.00	−3733.4	0.00	0.00	691	82	−3806.4	5.2453
1950	198.75	0.00	0.00	−4058.9	0.00	0.00	680	79	−4133.9	5.4677
2000	210.25	0.00	0.00	−4403.9	0.00	0.00	668	76	−4480.0	5.6943

.223 Remington 77-grain Sierra MatchKing bullet ballistics

Federal Gold Medal: .308 Win.: Sierra MatchKing BTHP

	SIGHT–IN DATA	FIELD DATA	POINT BLANK RANGE DATA			
Muzzle Velocity (fps)	2600	2600	TARGET	SIGHT–IN	POINT–BLANK	RANGE
Bullet Weight (grains)	168		HEIGHT	DISTANCE	HIGH	LOW
Sight Height (in)	1.5		(in)	(yd)	(yd)	(yd)
Sight-in Distance (yd)	100		2	152	92	171
Altitude (ft)	0	0	4	191	110	220
Temperature (deg F)	59	59	6	221	123	258
Pressure @ Sea Level (in Hg)	29.53	29.53	8	247	135	289
Relative Humidity (pct)	78.0	78.0	10	269	146	315
Wind Velocity (mph)		0				
Wind Angle (degrees)		180 (6.0 O'Clock)				
Incline Angle (degrees)		0				
Moving Target Speed (mph)		0.0				
Ballistic Coefficient	0.458					
MAX APPARENT TRAJECTORY (in)		0.2				

TARGET	SIGHT ADJUSTMENTS			TRAJECTORY VALUES					Drop from	ARRIVAL
DIST	ELEV	WIND	LEAD	ELEV	WIND	LEAD	VELOCITY	ENERGY	bore line	TIME
(yd)	MOA	MOA	MOA	(in)	(in)	(in)	(fps)	(ft-lb)	(in)	(sec)
0	0.00	0.00	0.00	−1.5	0.00	0.00	2600	2521	0.0	0.0000
50	0.00	0.00	0.00	−0.1	0.00	0.00	2504	2339	−0.6	0.0588
100	0.00	0.00	0.00	−0.0	0.00	0.00	2410	2166	−2.7	0.1198
150	1.00	0.00	0.00	−1.4	0.00	0.00	2318	2004	−6.2	0.1832
200	2.25	0.00	0.00	−4.5	0.00	0.00	2228	1852	−11.3	0.2492
250	3.50	0.00	0.00	−9.3	0.00	0.00	2146	1708	−18.2	0.3179
300	5.00	0.00	0.00	−16.0	0.00	0.00	2055	1574	−27.0	0.3894
350	6.75	0.00	0.00	−24.8	0.00	0.00	1971	1449	−37.9	0.4639
400	8.50	0.00	0.00	−35.7	0.00	0.00	1890	1332	−51.0	0.5416
450	10.50	0.00	0.00	−49.2	0.00	0.00	1810	1222	−66.5	0.6227
500	12.50	0.00	0.00	−65.3	0.00	0.00	1733	1120	−84.6	0.7073
550	14.50	0.00	0.00	−84.2	0.00	0.00	1658	1025	−105.7	0.7958
600	17.00	0.00	0.00	−106.3	0.00	0.00	1586	938	−129.9	0.8882
650	19.50	0.00	0.00	−131.9	0.00	0.00	1518	853	−157.6	0.9849
700	22.00	0.00	0.00	−161.3	0.00	0.00	1452	787	−189.0	1.0859
750	24.75	0.00	0.00	−194.8	0.00	0.00	1390	720	−224.6	1.1915
800	27.75	0.00	0.00	−232.7	0.00	0.00	1331	661	−264.6	1.3017
850	31.00	0.00	0.00	−275.6	0.00	0.00	1278	609	−309.6	1.4168
900	34.25	0.00	0.00	−323.8	0.00	0.00	1228	562	−359.9	1.5366
950	48.00	0.00	0.00	−377.7	0.00	0.00	1182	521	−415.9	1.6611
1000	41.75	0.00	0.00	−437.9	0.00	0.00	1142	486	−478.2	1.7903
1050	46.00	0.00	0.00	−504.8	0.00	0.00	1106	456	−547.1	1.9239
1100	50.25	0.00	0.00	−578.7	0.00	0.00	1075	431	−623.1	2.0615
1150	54.75	0.00	0.00	−660.2	0.00	0.00	1046	408	−706.7	2.2032
1200	59.75	0.00	0.00	−749.6	0.00	0.00	1020	388	−798.2	2.3485
1250	64.75	0.00	0.00	−847.4	0.00	0.00	997	371	−898.1	2.4974
1300	70.00	0.00	0.00	−954.0	0.00	0.00	975	355	−1006.7	2.6498
1350	75.75	0.00	0.00	−1069.7	0.00	0.00	956	341	−1124.6	2.8055
1400	81.50	0.00	0.00	−1194.9	0.00	0.00	937	328	−1251.9	2.9643
1450	87.50	0.00	0.00	−1330.1	0.00	0.00	920	316	−1389.2	3.1262
1500	94.00	0.00	0.00	−1475.6	0.00	0.00	905	305	−1536.8	3.2911
1550	100.50	0.00	0.00	−1631.8	0.00	0.00	890	295	−1695.1	3.4589
1600	107.25	0.00	0.00	−1799.1	0.00	0.00	876	286	−1864.4	3.6296
1650	114.50	0.00	0.00	−1977.8	0.00	0.00	862	277	−2045.2	3.8031
1700	121.75	0.00	0.00	−2168.3	0.00	0.00	849	269	−2237.8	3.9794
1750	129.50	0.00	0.00	−2371.0	0.00	0.00	836	260	−2442.6	4.1587
1800	137.25	0.00	0.00	−2586.3	0.00	0.00	823	253	−2659.9	4.3408
1850	145.25	0.00	0.00	−2814.6	0.00	0.00	811	245	−2890.4	4.5258
1900	153.50	0.00	0.00	−3056.3	0.00	0.00	799	238	−3134.2	4.7138
1950	162.25	0.00	0.00	−3311.9	0.00	0.00	788	232	−3391.9	4.9046
2000	171.00	0.00	0.00	−3581.8	0.00	0.00	777	225	−3663.8	5.0984

⌃ **.308 Winchester 168-grain Sierra MatchKing bullet ballistics**

Lapua Target: .308 Win.: Scenar

	SIGHT–IN DATA	FIELD DATA	POINT BLANK RANGE DATA			
Muzzle Velocity (fps)	2690	2690	TARGET	SIGHT–IN	POINT–BLANK	RANGE
Bullet Weight (grains)	167		HEIGHT	DISTANCE	HIGH	LOW
Sight Height (in)	1.5		(in)	(yd)	(yd)	(yd)
Sight-in Distance (yd)	100		2	158	96	178
Altitude (ft)	0	0	4	198	114	228
Temperature (deg F)	59	59	6	229	127	266
Pressure @ Sea Level (in Hg)	29.53	29.53	8	255	140	300
Relative Humidity (pct)	78.0	78.0	10	279	153	326
Wind Velocity (mph)		0				
Wind Angle (degrees)		180 (6.0 O'Clock)				
Incline Angle (degrees)		0				
Moving Target Speed (mph)		0.0				
Ballistic Coefficient	0.470					
MAX APPARENT TRAJECTORY (in)		0.1				

TARGET	SIGHT ADJUSTMENTS			TRAJECTORY VALUES					Drop from	ARRIVAL
DIST	ELEV	WIND	LEAD	ELEV	WIND	LEAD	VELOCITY	ENERGY	bore line	TIME
(yd)	MOA	MOA	MOA	(in)	(in)	(in)	(fps)	(ft-lb)	(in)	(sec)
0	0.00	0.00	0.00	−1.5	0.00	0.00	2690	2683	0.0	0.0000
50	0.25	0.00	0.00	−0.1	0.00	0.00	2595	2496	−0.6	0.0568
100	0.00	0.00	0.00	−0.0	0.00	0.00	2501	2320	−2.5	0.1156
150	0.75	0.00	0.00	−1.3	0.00	0.00	2410	2153	−5.8	0.1767
200	2.00	0.00	0.00	−4.1	0.00	0.00	2320	1995	−10.6	0.2401
250	3.25	0.00	0.00	−8.5	0.00	0.00	2232	1847	−16.9	0.3060
300	4.75	0.00	0.00	−14.6	0.00	0.00	2146	1708	−25.1	0.3745
350	6.25	0.00	0.00	−22.6	0.00	0.00	2063	1578	−35.1	0.4458
400	7.75	0.00	0.00	−32.7	0.00	0.00	1981	1455	−47.2	0.5199
450	9.50	0.00	0.00	−45.0	0.00	0.00	1902	1341	−61.4	0.5972
500	11.50	0.00	0.00	−59.6	0.00	0.00	1824	1233	−78.1	0.6777
550	13.25	0.00	0.00	−76.9	0.00	0.00	1748	1133	−97.4	0.7617
600	15.50	0.00	0.00	−97.0	0.00	0.00	1675	1040	−119.5	0.8493
650	17.75	0.00	0.00	−120.3	0.00	0.00	1604	954	−144.7	0.9408
700	20.00	0.00	0.00	−146.9	0.00	0.00	1536	875	−173.4	1.0364
750	22.50	0.00	0.00	−177.2	0.00	0.00	1472	803	−205.6	1.1361
800	25.50	0.00	0.00	−211.5	0.00	0.00	1410	737	−241.9	1.2402
850	28.00	0.00	0.00	−250.1	0.00	0.00	1351	677	−282.6	1.3489
900	31.25	0.00	0.00	−293.6	0.00	0.00	1297	624	−328.0	1.4622
950	34.50	0.00	0.00	−342.1	0.00	0.00	1247	577	−378.6	1.5801
1000	37.75	0.00	0.00	−396.3	0.00	0.00	1201	535	−434.8	1.7026
1050	41.50	0.00	0.00	−456.5	0.00	0.00	1159	498	−497.0	1.8298
1100	45.50	0.00	0.00	−523.2	0.00	0.00	1123	467	−565.6	1.9614
1150	49.50	0.00	0.00	−596.7	0.00	0.00	1090	441	−641.2	2.0971
1200	54.00	0.00	0.00	−677.6	0.00	0.00	1061	417	−724.0	2.2367
1250	58.50	0.00	0.00	−766.2	0.00	0.00	1034	397	−814.6	2.3800
1300	63.50	0.00	0.00	−862.9	0.00	0.00	1010	379	−913.4	2.5269
1350	68.50	0.00	0.00	−968.2	0.00	0.00	988	362	−1020.6	2.6773
1400	73.75	0.00	0.00	−1082.4	0.00	0.00	968	348	−1136.8	2.8309
1450	79.50	0.00	0.00	−1205.9	0.00	0.00	949	334	−1262.3	2.9877
1500	85.25	0.00	0.00	−1339.0	0.00	0.00	932	322	−1397.5	3.1476
1550	91.25	0.00	0.00	−1482.3	0.00	0.00	916	311	−1542.7	3.3105
1600	97.75	0.00	0.00	−1635.9	0.00	0.00	901	301	−1698.3	3.4762
1650	104.25	0.00	0.00	−1800.4	0.00	0.00	887	291	−1864.8	3.6447
1700	111.00	0.00	0.00	−1975.9	0.00	0.00	873	282	−2042.4	3.8160
1750	118.00	0.00	0.00	−2163.0	0.00	0.00	859	274	−2231.5	3.9901
1800	125.25	0.00	0.00	−2362.0	0.00	0.00	847	266	−2432.4	4.1670
1850	132.75	0.00	0.00	−2573.3	0.00	0.00	834	258	−2645.7	4.3467
1900	140.50	0.00	0.00	−2797.2	0.00	0.00	822	251	−2871.6	4.5292
1950	148.58	0.00	0.00	−3034.2	0.00	0.00	810	244	−3110.6	4.7146
2000	156.75	0.00	0.00	−3284.7	0.00	0.00	799	237	−3363.1	4.9027

≫ **.308 Winchester 167-grain Lapua Scenar bullet ballistics**

RECOMMENDED READING

Basham, Lanny. *With Winning in Mind, Third Edition*. Mental Management Systems (2011).

Written by a former Olympic gold medalist and world champion shooter, this is an excellent book on proper mind-set applied by a champion. One of the best books on mental management I've ever read.

Blehm, Eric. *Fearless: The Undaunted Courage and Ultimate Sacrifice of Navy SEAL Team Six Operator Adam Brown*. Waterbrook Press (2012).

A great read, this SEAL memoir also has some amazing insights on sniper training regarding right- and left-eye-dominant shooting. The subject of the book, SEAL Adam Brown, goes through sniper school with one eye and has to learn to shoot left-handed (he lost his dominant eye in a training accident).

Chivers, C. J. *The Gun*. Simon & Schuster (2010).

While this book deals with the ubiquitous AK-47, it also covers some great historical content with the advent of the U.S. military's modern semiautomatic rifle.

Henderson, Charles. *Marine Sniper: 93 Confirmed Kills*. Berkley (2001).

What collection would be complete without the original USMC master sniper himself, Carlos Hathcock? A great book on everything USMC sniping. The information in this book is proof that a sniper is one of the most deadly tools in a commander's kit.

Kyle, Chris, with Scott McEwen and Jim DeFelice. *American Sniper: The Autobiography of the Most Lethal Sniper in U.S. Military History*. William Morrow (2012).

Chris is a friend of mine and his book is a must-read for anyone interested in the subject of sniping. He has some excellent examples of practical application on the battlefield.

Luttrell, Marcus. *Lone Survivor: The Eyewitness Account of Operation Redwing and the Lost Heroes of SEAL Team 10*. Little, Brown and Company (2009).

Marcus was one of Brandon's students during his tenure as SEAL Sniper Course Manager. *Lone Survivor* is an excellent read and an example of how sniper skills can save your life when it comes to stealth and concealment on the battlefield.

Maylor, Rob, with Robert Macklin. *Sniper Elite: The World of a Top Special Forces Marksman*. St. Martin's Griffin (2012).

This book shares Rob's personal experiences as an Australian SAS sniper on the battlefield and also some insight into training. An excellent read; we highly recommend grabbing his book.

Rinker, Robert A. *Understanding Firearm Ballistics*. Mulberry House Publishing Co (1999).

This must-have for any sniper covers all things ballistics.

Tubbs, David. *The Rifle Shooter*. Zedicker Publishing (2003).

If you want a book that covers down and dirty fundamentals of marksmanship, look no further. A world championship shooter, David covers the fundamentals in great detail.

Webb, Brandon, with John David Mann. *The Red Circle: My Life in the Navy SEAL Sniper Corps and How I Trained America's Deadliest Marksmen*. St. Martin's Press (2012).

I'm recommending my own book because it gives you a great inside look at what goes into training a modern-day sniper.

ACKNOWLEDGMENTS

I had no idea that I'd be adding the finishing touches to this book without my shooting partner Glen.

Tony Lyons at Skyhorse deserves a big thank-you for making this book into something Glen would be proud of. When I called after Glen died in Libya and said, "This book needs to be a hardback and the print quality better be good," Tony agreed on the spot and made it happen.

For those of you who don't know, I wrote this book with Glen before my memoir, *The Red Circle*. We both had no idea what we were getting into when we came up with the idea for this book. Many people have helped along the way and deserve thanks, appreciation, and recognition for their time and effort.

Our good friend Dr. Curtis Prejean provided incredible content for us. He is a virtual encyclopedia of knowledge concerning weapons and ballistics, and an all-around great guy to boot.

Ted Rush was an incredible help with keeping people on deadline, myself included. Glen is smiling down on all of us for doing this one right. Thank you Ted! We can still have that beer we discussed, but we're down one good man. We'll just have a full glass at the table in honor of Glen.

Thanks to my *Red Circle* coauthor John David Mann for jumping into the fray and lending his wordsmithing skills to this revised edition of *The Twenty-First-Century Sniper*. I can't thank you enough, John. You've taught me so much about being a good writer and I thank you sincerely for all your help. *Lost Heroes* and *Navy SEAL Sniper* wouldn't be the same books without your help.

Thanks to Matt Johnson for getting out of his banker's chair to find some great old photos for Glen and me to use in the book, and to Scott Tyler for his stories and photos that he was willing to donate to the cause.

To Marc Halcon of American Shooting Center in San Diego for his network, his knowledge, and access to all his great range.

Thank you to the SOFREP.com community for all their support.

I want to thank Glen's family for all their support. His sister, Kate, and brother, Greg, are such wonderful human beings.

To his large circle of friends, I want to say this: If you were a friend to Glen, you are a friend of mine, and I'll always have a guest room available and cold beer on standby.

Lastly I want to acknowledge the tremendous work that Glen put into this book; it would have never been written without his help. He was a best friend, a brother, and one of the finest human beings I've ever known. See you on the other side, Bub.